Regenerating Essential Goods and Services in Urban Landscapes

How do we provide for and nurture millions of people without destroying the planet in the process?

Author Doug Kent, an environmental specialist, believes a vital element in the solution is recognizing that urban landscapes are an essential partner in everyone's wellbeing. He argues that urban landscapes can and must work harder.

Urban landscapes can provide part of our energy needs, help cool our buildings and public spaces, help us make the most of our precious water. They can also help combat air pollution and reduce the likelihood of allergies and asthma. They can provide landscape materials and even contribute to our timber supply. Doug also advocates turning landscapes into a food source, and/or a perfumery, pharmacy, soap shop, or craft store.

Doug has over 12 years of research in this book. He has spent years doing literature reviews, and many more years concocting, consuming, crafting, distilling, propagating, retting, sawing, sowing, and weaving its many recommendations. He has also travelled the length and width of California many times to interview the people and businesses already doing this incredible work.

Regenerating Essential Goods and Services is not a manifesto. It is a user's manual. You are the creative and energetic force that will ultimately drive sustainability and regeneration. Let's go.

Doug Kent is an author, activist, and educator in ecological land management. He has been exploring the dynamics and ecology of human sustainability for decades. He started gardening in 1979, began pursuing environmental remedies in 1988, became a landscape designer/contractor in 1995, embraced urban regeneration in 2003, and started teaching land management in 2005. Currently, he teaches at the Center for Regenerative Studies at California Polytechnic University, Pomona, UCLA Extension Horticulture, and USC Landscape Architecture and Urbanism.

Regenerating Essential Goods and Services in Urban Landscapes

Sustainability Through Ecological Design

Douglas Kent

Routledge
Taylor & Francis Group

NEW YORK AND LONDON

Designed cover image: © Douglas Kent

First published 2024
by Routledge
605 Third Avenue, New York, NY 10158

and by Routledge
4 Park Square, Milton Park, Abingdon, Oxon, OX14 4RN

*Routledge is an imprint of the Taylor & Francis Group, an informa
business*

© 2024 Douglas Kent

ISBN: 9781032439860 (hbk)
ISBN: 9781032439853 (pbk)
ISBN: 9781003369752 (ebk)

DOI: 10.4324/9781003369752

Typeset in Times New Roman and Futura Std
by Deanta Global Publishing Services, Chennai, India

Who is more important than the reader? Nobody, that's who.

This work is dedicated to you. It is for the curious, the engaged, and the big-hearted. It is for all who appreciate both the fragility of our planet and its inspiring wonderment.

I am proud to be working alongside you.

Annotated Table of Contents

Figures

Tables

About the Author

Doug Kent is an author, activist, and educator in ecological land management. He has been exploring the dynamics and ecology of human sustainability for decades. He started gardening in 1979, began pursuing environmental remedies in 1988, became a landscape designer/contractor in 1995, embraced urban regeneration in 2003, and started teaching land management in 2005.

Doug is the author of eight environmental land management books, including *Firescaping: Protecting your home with a fire-resistant landscape* (Wilderness Press 2019, 2nd Ed.); *Foraging Southern California: 118 nutritious, tasty and abundant foods* (Adventure Publication); *California Friendly: A Maintenance Guide for Landscapers, Gardeners and Land Managers* (MUNI 2017); *Ocean Friendly Gardens: A how-to gardening guide to help restore a healthy coast and ocean* (Surfrider Foundation 2009); and *A New Era of Gardening: A book on gardening for oxygen and a healthier atmosphere* (Garden Shed Productions 2001). He has also written articles for some of the state's largest newspapers, including the *Los Angeles Times*, *Orange County Register*, *Marin Independent Journal*, and *Press Democrat*.

More than anything, Doug loves to teach. He has been teaching land management for over 16 years. Currently, he teaches at the Center for Regenerative Studies at California Polytechnic University, Pomona, UCLA Extension Horticulture, and USC Landscape Architecture and Urbanism. He has also taught sustainable landscaping classes for California's largest water and power providers, conservation agencies, and environmental non-profits since 2009.

His work can be viewed at www.anfractus.com and he can be reached at newair@mindspring.com.

Acknowledgments

Writing sustainable landscaping books is not a job for someone who wants to make a quick profit. It's for someone who wants to change the world for the better, which I unapologetically do. I also happen to find the process an epic adventure and joy. I feel lucky to have had so many supporters who appreciate the art, research, and time required to produce a substantial book like this. These heroes include:

Angel City Lumber, AdventureKEEN, Air Burners Inc., Breathe Southern California, Center for Regenerative Studies, Custom Landscapes, Debbie Dunne (fantastic twin), Fibershed, Humboldt Sawmill Company, Juan Araya, Moonwater Farm, ReScape California, Richmond Greenwaste Recycle Yard and Millworks, Routledge, Sierra Club members, Slane Whiskey, Stone Brewing, Sustainable Claremont, Wilderness Press, USC Landscape Architecture+Urbanism, and many family members and friends have all given energy and time, and more.

That said, there are three people that deserve more than a mention. These were the propellants.

Sharon Cohoon, editor: Sharon leapt into this work like a teenager would a clear lake on a hot summer day. She did far more than scrub and sooth, she edited with a drive and voice that was in harmony with mine. I am proud of our work.

Kathleen Blakistone, co-owner of Moonwater Farm: Kathleen has rallied this book for years. All the time we spent in classes, in her gardens, and on incredible adventures reverberate throughout this work. Kathleen is daring, hands-on, and a powerful advocate for ecological and equitable health. There are many photos of Moonwater in this book.

Lyle Center for Regenerative Studies, California Polytechnic University, Pomona: Since 2004 the staff, faculty, and students have given me the opportunity to study, explore, harvest, cook, craft, push, write, teach, and stumble. So much of this book has been battle tested at the Center. I am forever grateful to be able to roam 16 acres of good intention, and for so long.

Chapter 1

Introduction

How do we do it?

How do we care for millions of people without destroying the planet? How do we, as environmentalists, land-tenders, and city influencers enable and nurture an urban ecology that cares for all?

The answer, as this book addresses, lies in creating an urban ecology that supports humans. We design, maintain, and honor our landscapes as life sustaining entities. We support reciprocal systems that generate goods and services, health and wellbeing. And as ecologists, we cultivate the dynamic relationships that sustain this bounty. We value microorganisms and worms, birds and pollinators, lizards and mesopredators as much as we do the goodness of plants, recognizing their equally vital contribution to healthy urban ecosystems.

We can do all this. If we apply our creativity, energy, and time we can ignite an ecosystem that will provide for us as well as them. Urban landscapes are ideally suited for regenerating essential goods and services because of these three reasons:

- **Opportunity**: Urban areas have more money, time and resources to devote to their landscapes than agricultural or rural areas. We are rich in capability. Urban landscapes are also underutilized. Aesthetics, safety, resource conservation, and attracting pollinators are the current tasks. We can provide all those things, and, as this book argues, so much more. We are rich in opportunity.
- **Species Loss**: Urban areas cannot truly tackle localized species loss. It is not because there are not enough native plants, but because there are too many people, cars, heat, hard surfaces, dogs, cats, light, noises, gaseous and particulate pollutants, and chemically different soils and waters. Urban areas also lack disturbances, such as fires, floods and apex predators that have historically defined natural habitats. Because of all these factors human nature creates a nature that is not suited for native niche species. However, urban areas can protect native species from being eradicated elsewhere by keeping our basic needs from overrunning the landscapes where they reside.
- **Half Earth**: The surest way to slow or stop climate change and specie loss is to set aside half the planet for native and natural endeavors:[1] Half of Earth is left to

DOI: 10.4324/9781003369752-1

run wild. The other half is dedicated to human needs. The only way this idea will work is if urban areas start shouldering the needs of their consumptive members. We need to stop overrunning distant landscapes and start creating deeper and richer connections to our own.

The most dynamic part of this work is its potential change to the perceptions and practices of landscape management. Design becomes verifiable. Maintenance becomes a science. And utilizing vegetation becomes an art form. Residential gardeners, landscape architects, and city planners work in concert to grow health and wellbeing, and, consequently, reduce species loss and climate change.

The most exciting part of this work is creating deeper connections to urban landscapes. We encourage and participate in reciprocal relationships. We nurture an urban land ethic that supports resilience, persistence, and replication. We show up to graciously give and receive.

How do we do it? Let's dig in and find out.

See you in a garden.

Figure 1.1 **We can do it. We can create an urban ecosystem that cares for us as well as them. Let's dig in!**

Note

1 Wilson, Edward. *Half-Earth: Our Planet's Fight for Life*. Liveright Publishing Corporation. April 2017.

Chapter 2

Craft Items and Textiles

If you have ever harvested materials for craft items or produced textiles from nature you know the joy, satisfaction, and sense of accomplishment the process produces. The colors are more like Mother Earth, the fiber creations more of a keepsake, and the ceremonial items are more meaningful. The process involves harvesting, processing, and concocting, all of which is dynamic, physical, and rewarding.

But personal satisfaction isn't the only benefit of generating craft materials and textiles. It diverts greenwaste from the waste stream. It can be healthier for the environment than industrial processes. It can be good for personal health. And it can be a financially rewarding profession.

Creating landscapes that produce craft items is ideally suited for urban areas for three reasons. First, the crop never spoils, and it doesn't attract pests like food does. Second, craft/textiles are specialty crops, which is something industrial processes struggle to produce. Third, it fits right in with many of the other endeavors in this book, such as energy, food, public health, thermal comfort, and water capture.

Included below is an overview of the plants and practices for creating dyes, fibers for cordage and weaving, paper, and religious and seasonal items.

Quick Overview of Craft Items and Textiles[1]

Uses: Dyes are used on cordage, fabrics, paper, hide and wood. Cordage is used for thread, twine, rope and weaving, for such things as baskets and mats. Homemade paper is used mostly for cards, notes and seed props (seeds are imbedded in the paper). Religious/seasonal items are used for celebrations, holidays, and remembrances.

Costs: Low to medium.

DOI: 10.4324/9781003369752-2

Difficulty: None of the processes below are difficult. Humans have been doing all of them for thousands of years. The hurdle is devoting the time and attention these pursuits deserve.

Best Scale: Landscapes planted with craft items and utilized as such are ideal in many urban landscapes, particularly around K-12 schools, public parks, college campuses, art schools, and nature centers that desire more community participation.

ERoEI: The calculations and research needed to thoroughly address ERoEIs in this area would be complex and exhausting, and as of yet they have not been calculated. However,

Figure 2.1   Dye pie Pictured are common urban dyes. From the top and clockwise are redbud bark, elderberry twigs, cochineal, pomegranate skins, pinecone, coyote brush tops, mint leaves, Chinese tallow tree leaves, Catalina cherry leaves, fennel leaves, alder bark, *Eucalyptus* **bark, red onion skins, avocado skins. and arroyo willow bark.**

creating textiles you use on a regular basis will supplant some industrial processes. And, while not all industrial textiles are energy-drenched and polluting, most assuredly are. Urban dwellers can create more earth-friendly products because they are not necessarily motivated by a bottom line.

Pros: Nearly every step of creating these items is good for your health. Whether you are growing, harvesting, processing, or using the end products, at every stage you are handling natural and mostly non-toxic material.

Cons: Time and attention are the two largest drawbacks for any of the endeavors in this book. Fossil fuels and toxic processes are replaced with our energy, time, and creativity.

Dyes

Producing colors that captivate and engage, and fragrances that linger and inspire, the process of working with natural dyes is richly rewarding. The natural pigments below can be used on a variety of materials, including cloth (animal/plant), paper, hide, and wood.

What follows below is a flash introduction into making dyes. Warning: Do not cook with the pots and utensils used for dying. Some of the processes below become caustic.

Nomenclature

Adjective Dyes: A dye that is not colorfast and will lose its color without a fixative or mordant.

Animal Fiber: Fibers made from animals. The most common are cashmere (goat), feathers, rabbit, silk (worm) and wool (sheep). These fibers differ from plant fibers because they have protein.

Bast: The inner bark of a plant and the part of its vascular system called the phloem. Bast fibers are prized.

Dye Bath: The final solution—a crafted blend of materials, additives, and temperature.

Colorfast: Dyes that hold their colors against sunlight and washing.

Fugitive: Dye prone to fading, more so when exposed to sunlight or washing.

Lightfast: Dyes that hold their color when exposed to light, particularly the sun.

Mordant: A substance that helps bind a dye to a fiber. Mordants can alter the color of a dye. Mordants are such things as alum, iron, and plant parts high in tannins, such as oak galls.

Prepared For Dyeing (PFD): Fibers that have been thoroughly cleaned, or scoured, and are ready for the dye bath. The process typically involves boiling water with something either highly acidic, like vinegar, or highly alkaline, like soda ash.

Plant Fibers: Fibers derived from plants and high in cellulose. Cellulose is a complex carbohydrate and the foundational structure in plant cell walls. On average 33% of all plant matter is cellulose. Plants high in cellulose, such as cotton at 90%, are fiber crops.

pH: A scale that represents how alkaline or acidic a solution is. The pH of a dye bath can alter the color. Baking soda will make it more alkaline and vinegar more acidic.

Soda Ash: A white powder or granule containing more than 99% sodium carbonate.

Soda ash forms a strong alkaline water solution which reacts with the fiber allowing the dye to make a permanent connection. Do not use soda ash on protein fibers, such as wool.

Stain: A stain is more fleeting when compared to a dye. The color does not penetrate as deeply and is not as long lasting as a dye. Stains tend to be topical and less deep, which makes a vibrant color for a short period of time.

Substantive Dyes: These are dyes that are colorfast and durable. Most of these dyes have tannic acid, which is a natural mordant. Three of the most notable are avocado skins, redbud bark, and willow bark.

Tannins: Astringent and bitter compounds found in many parts of some plants. Parts high in tannins can include bark, fruit skins, galls, leaves, and seeds. Tannic acid is a natural mordant.

COMMON URBAN DYE PLANTS

Legend
C: Cultivated
N: Naturally found

Alder, *Alnus* spp.: Bast makes brown, orange, red. C, N

Ash, foothill, *Fraxinus dipetata*: Ground charcoal from wood makes black. C, N

Avocado, skins and stones make pink, reddish pink, rose and rust red. C

Barberry (mountain grape), *Berberis aquifolium* and *B. pinnata*: Leaves make a greenish color, bast and roots yellow and brown, and berries blue, pink, purple, green. Berries are a stain, not a dye. C, N

Bean, black, *Phaseolus vulgaris*: Whole bean makes blue, green. Produces a stain, not a dye. C

Beet, *Beta vulgaris*: Roots make pink, red. C

Blackberry, *Rubus*: Berries make maroon, pink, purple. Produces a stain, not a dye. C, N

Figure 2.2 **Avocado skins and pits make a rich colorfast dye. The darker color, to the right, is made from the skins, and the lighter color, to the left, is from the pits.**

Blueberry, *Cyanococcus*: Berries make blue, pink, purple. Produces a stain, not a dye. C

Broom, French, *Genista monspessulana*: Leaves and stocks make a greenish yellow. C, N

Broom, Scotch, *Cytisus scoparius*: Leaves and stocks make yellow, beige, light mustard. C, N

Burnweed, American, *Erechtites hieraciifolius*: Leaves make blue. N

Cabbage, red, *Brassica oleracea*: Cabbage is finely diced to make blue, green, pink, purple. Soda ash and alum will influence color. Cabbage is a stain, not a dye. C

Calendula, *Calendula officinalis*: Flower heads make gold, yellow. Calendula make a stain, not a dye. C, N

Ceanothus, *Ceanothus* spp.: Roots make a reddish color. C, N

Chastetree, *Vitex agnus-castus*: leaves make butter to mustard yellow and algae green with alum. C

Cherry, *Prunus* spp.: Fruit, leaves and root make green, pink, red, tan, yellow. C, N

Chinese tallow tree, *Triadica sebifera*: Leaves make sunny mustard yellow. C, N

Cochineal: Found on *Opuntia*, bug makes crimson, pink, purple. C, N

Coral bells, *Heuchera* spp.: Root is used as a mordant. C, N

Coreopsis or golden tickseed, *Coreopsis tinctoria*: Flowers make a vibrant orange to yellow. C, N

Coyote brush, *Baccharis pilularis*: Leaves and stems make yellow. C, N

Dock and wild rhubarb, *Rumex* spp: Flower stalks make grey to yellow; leaves red, yellow; roots cinnamon, gold, yellow. C, N

Elderberry, blue, *Sambucus nigra caerulea (S. mexicana)*: Fruit makes grey to purple, small stems and twigs make orange, pink/tan to yellow. Berries produce a stain, not a dye. C, N

Eucalyptus: Leaves make brown, orange, yellow; bark makes brown, pink. C, F, W

Fennel, *Foeniculum vulgare*: Leaves make yellow, lime green, dark grey green with iron. Requires a large mass of leaves. C, N

Fig, *Ficus carica*: Leaves make a golden yellow, mustard. C, N

Filaree, *Erodium cicutarium*: Leaves make pale yellow, green. N

Goldenrod, *Solidago* spp.: Flowers make light brown to light sunny yellow. C, N

Hazelnut, beaked, *Corylus cornuta*: Roots make blue. C, N

Henna, *Lawsonia inermis*: Leaves make a variety of colors, from blond and orange to reddish tan and dark brown. C

Horsetail, *Equisetum arvense*: Stalks make yellow. C, N

Indigo, *Indigofera tinctoria* (tropical indigo) and *Persicaria tinctoria* (temperate indigo): Leaves make blue. C

Ironwood, island, *Lyonothamnus* spp.: Bark makes black. C

Lambsquarter and **goosefoot**, *Chenopodium album* and *C. murale*: Entire plant makes beige, gold, greenish, and yellow. C, N

Lavender, *Lavandula* spp.: Flowers make brown, pink, yellow. C

Lichen, many Genera: Dried tops make a range of colors, from yellow to brown, red to purple. N

Lilac, *Syringa vulgaris*: Flowers may make green, leaves greenish brown, twigs yellow and orange. C, N (limited)

Madder, *Rubia tinctoria*: Roots make pink to red. C

Madrone, *Arbutus* spp.: Bark makes reddish to pink. C, N

Madrono, *Arbutus menziesii*: Bark makes a brown. C, N

Maple, Japanese, *Acer palmatum*: Leaves make pink, purple. C, N

Marigold, *Tagetes* spp.: Flowers make sunny to burnt yellow. More of a stain than a dye. C, N

Mint, *Mentha* spp.: Leaves make greenish, grey, yellow. Every variety has its own color. C, N

Mulberries, *Morus* spp.: Berries make pink, purple, although it is a stain, not a dye. C, N

Mullein, common, *Verbascum thapsus*: Flowers make yellow, greenish, brown. C, N

Oak, *Quercus* spp.: Acorns make brown, galls blackish. Galls are a mordant. C, N

Onion, *Allium*: Red or yellow skins make gold, yellow olive, orange, red, yellow. C

Pearly everlasting, *Anaphalis margaritacea*: Use flowers, leaves, stems for yellow, greenish, brown. C, N

Pine, *Pinus* spp.: Cones make pink. C, N

Poinsettia, *Euphorbia pulcherrima*: Bracts make a khaki green to pink that is almost red. C, N

Pomegranate: Skins make pink, purple, yellow, and green when a mordant is added. Makes a stain, not a dye. C, N

Pokeweed, *Phytolacca americana*: Berries make a magenta to purple. Makes a stain, not a dye. C, N

Queen Anne's lace (wild carrot), *Daucus carota*: Leaves make light yellow and mustard. C, N

Rabbitbush, *Ericameria nauseosa*: Flower heads make gold, greenish, yellow. C, N

Redbud, *Cercis* spp.: Bark makes rose tan. The outer bark produces a deeper red; bast is browner. C, N

Sagebrush, *Artemisia* spp.: Leaves and sprigs make gold, yellow. C, N

Saffron, *Crocus sativus*: Stigmas make yellow, orange and mustard. C

Sourgrass, *Oxalis pes-caprae*: Flowers make orange, yellow. A stain not a dye. N

Stinging nettle, *Urtica* spp.: Leaves make greenish, yellow; roots yellow. C, N

Sumac, *Rhus* spp.: Berry heads make grey, purple, red; roots brown, yellow. C, N

Sweetgum, *Liquidambar* spp.: Leaves make yellow to dull reddish and bark beige to a light red/pink. C, N

Sycamore and planetree, *Platanus* spp.: Outer bark makes brownish red. C, N

Tansy, *Tanacetum vulgare*: Flowers make yellow, leaves greenish. C, N

Tree of heaven, *Ailanthus altissima*: Leaves make mustard, yellow and are high in tannins. N

Turmeric, *Curcuma* spp.: Tuber makes rich golden orange to yellow. C

Viper's bugloss, *Echium vulgare*: Root makes red. C, N

Figure 2.3   **Tree of heaven is a common weedy tree throughout much of North America. The clothes above were dyed using its leaves and the upper and richer cloth had a mordant, aluminum sulfate, added to the dye bath. The tree is also a fiber crop and can be made into pesticides.**

Walnut, black, *Juglans nigra*: Hulls make blackish, brown; leaves yellow. C, N

Wattle, *Acacia* spp.: Bark, flowers, and seedpods produce yellow to greenish. C, N

Weld (dyer's rocket), *Reseda luteola*: Leaves, flowers and seeds make sunny yellow to dark mustard. C, N

Willow, *Salix* spp.: Bark and twigs make burgundy to red brown. Outer bark produces a darker, bloodier red; the bast makes a light red/pink. Leaves produce yellow. N

Woad: *Isatis tinctoria*: Leaves make a blue. C, N

Yarrow, *Achillea millefolium*: Flowers, leaves, twigs make brown, greenish, yellow. C, N

*Importantly, thoroughly clean/scour your material before putting it into the dye bath. The dye bath can become tainted with unwanted bugs, debris, dirt, and dust.

Fibers for Cordage, Containers and Woven Creations

Urban environments are rich with materials that can be made into practical and stunning creations. Whether twine for wrapping or blades for baskets, fiber products can satisfy many everyday needs.

Fibers are harvested from various parts of plants:

- **Bast**: Inner bark from such plants as cedar, milkweed, and willow.
- **Fruits**: Fibers from fruits include cotton, kapok and the outer fibers of coconut (coir).
- **Leaves**: Fibers from leaves come from plants like cattail, grass, palm, and yucca.
- **Roots**: Fibers from plants like fir, grape, and willow.
- **Stalks**: Fibers from stalks generally need a lot of processing and include bamboo, grape, flax, hemp, and nettle.

Nomenclature

Animal Fibers: Fibers made from animals. The most common are cashmere, feathers, rabbit, silk and wool. These fibers differ from plant fibers because they have protein. They take colors and processes differently.

Bast: The inner bark of a plant and the part of its vascular system called the phloem. Bast fibers are prized for their fineness.

Braiding: Interweaving three or more strands together. There are several ways to braid. Also called plaiting.

Figure 2.4 Pictured is twine created from plants commonly found in urban areas. From top left and right are *Arundo*, dandelion, arroyo willow, narrow leaf milkweed, Spanish bayonet, and Mexican fan palm.

Figure 2.5 Woven from sheep raised on a nearby farm and dyed with indigo grown onsite, this sweater is not only hyper local, but also completely compostable. The rolls of cotton cloth were also grown on a nearby farm. Picture taken at Fibershed Headquarters in Pt. Reyes, CA.

Cane: A thick and more rigid material used as a stake (the bones of a container).

Coating: Natural creations are prone to decay and can be preserved with a variety of coatings, such as animal fats, plant oils and tar.

Cord: Twisted fibers that are thinner than rope, but thicker than twine.

Cordage: The accumulation of twisting strands that creates twine, cord and rope. At its most basic, creating cordage involves twisting long strands of anything together. The more strands and the more twists, the greater the strength.

Drying: Some plant material, like grasses, should be mostly dry before twisted into cordage because they shrink if not, causing your creation to unwind. Other material, like fronds, should be slightly moist to avoid cracking.

Extracting: Some plants, like palms, can be used directly, others need to be processed, extracting the long fibers from the other matter. Many of the plants listed below need processing. The goal of extraction is to separate cellulose from lignin and pectin. Each material requires a slightly different approach. Some of the methods are beating, freezing, heckling, immersing, pounding, retting, scrapping, scutching, and stripping. Extracting is often called Dressing.

Filament: The building block of cordage. It is a long and single sinuous strand. From the filament grows thread, twine, string, and rope.

Plant Fibers: A cellulose rich fiber made from plants. Field fibers include agave, grass, cattail, iris, reeds, stinging nettle, willow, and yucca. Grown fibers include bamboo, cotton, flax and hemp.

Retting: Water and/or microorganisms are employed to separate the useful fibers in a plant from all the other, strength-reducing, tissue. There are many ways to ret a plant and boiling in highly alkaline water is one of the most common.

Rope: The strongest and thickest of your twisted cordage.

Scutching: The beating and combing of retted material to further refine the long useful fibers. There are many ways to scutch material.

Stake: The bones of the basket or container

Thread: The smallest and weakest of cordage and comprised of one or more twisted filaments.

Twine: The word means to twist together. The product, twine, is stronger than string, but weaker than cord and rope. Twine is comprised of two or more strands twisted together, unlike a braid that has three or more.

Weaver: Light material woven around the stakes to create a basket or mat.

Weave, Types: There are many patterns of weave, the most common being randing, pairing and waling.

Figure 2.6 This beehive was woven from the blades of grasses and reeds collected within 100 feet of it. It is effective, functional and completely compostable. Picture taken at the Marin Art and Garden Center, Ross, CA.

COMMON FIBER PLANTS

Legend
C: Cultivated
F: Naturally found

Agave, *Agave* spp.: Requires scraping and pulling to get fibers. Leaves can be lightly cooked before scraping to make the task easier. Beware of its juice, which is an allergen and irritant (although it can be cooked out). Avoid plump-leafed varieties. C, N

Arundo, *Arundo donax*: Stalks require pounding and pulling to get fibers. Leaves require scrapping. Used for weaving and rough cordage. N

Bamboo, *Bambusoideae* spp.: Requires pounding and pulling to get fibers. Cordage is one of the softest. C, N

Bulrush and tule, *Bolboschoenus* spp. and *Schoenoplectus* spp.: Requires shredding and pulling. Used more for weaving than cordage. C, N

Cattail, *Typha* spp.: Involves splitting and tearing. Used more for weaving than cordage. N

Cedar, *Cedrus* spp. and *Thuja plicata* (western red): The bast makes fine cordage. Preparation involves cutting away the outer bark, pounding, and pulling the fine inner strands. Roots can be pealed and split for fiber too. C, N

Cotton, *Gossypium barbadense* and *G. hirsutum*: Easy to grow in areas with at least 180 growing days. In some states, like California, cotton can only be legally grown from March to October, the reason being is that if they grow year round they can breed pests that threaten agricultural interests. Seeds also make oil. C

Dandelion, *Taraxacum* spp.: A weed throughout the U.S. Harvest the longest leaves and flower stems, completely dry them, slightly moisten them, and start twining.

Elderberry, *Sambucus* spp.: Bast is used for clothes and rope. C, N

Flax, *Linum usitatissimum*: Bast makes linen and is prized for its softness, colorfastness, and strength. *L. grandiflorum*, scarlet flax, a weed, is also used. Pressing and scutching are required for fibers. C, N

Grape, *Vitis* spp. Pounding and stripping the long roots are needed to create rope. Bark needs a thorough soaking for string. C, N

Grass: While all grass can make cordage, the longer the blade, the easier it is to make and the stronger it is. Some of the longer grasses include deer, fountain, lemon, Lyme, *Miscanthus*, needle, oat, rye and switch grasses. C, N

Hemp: *Cannabis sativa*: Stalks are harvested, retted, scutched (combed), and used for a wide variety of everyday items. Hemp produces several types of fibers. C

Indian hemp (dogbane), *Apocynum cannabinum*. A perennial that dies back in the heat, grows in gravely/sandy soils, and can be found throughout the U.S. (desert and mountaintops excluded). Harvest dried stalks, crush, and pull the fine outer fibers from the inner core. C, N

Iris **spp.**: Preparation of the long leaves involves scrapping and stripping. C, N

Kenaf, *Hibiscus cannabinus*:. The bast and inner core of the long stalks are harvested. The bast produces a course fiber, the core a fine fiber. The bast is split from the core and both are retted, either biologically or chemically.

Kapok, *Ceiba pentandra*: Soft fibers are pulled from seedpods. Because the fiber is short and brittle it is better used as stuffing or as an addition to yarn fibers.

Milkweed, *Asclepias* spp. Not the easiest fiber to extract, but it is beautiful. The outer layer of the stem must be removed from the bast and it can either be lightly scrapped off or crushed and scrunched off when the stem is dry. The narrow milkweed (*A. fascicularis*) produces a fine fiber with a luxurious sheen. The balloonplant or family jewels milkweed (*Gomphocarpus physocarpa*) produces a strong white fiber with the texture of monofilament. C, N

New Zealand flax, *Phormium tenax*: Scrap leaves and either use or continue processing by beating, rubbing and/or retting. Leaves make ropes and clothes, mats and slings. C

Palm: The leaves of palms have long been split for cordage. Mexican fan palm (*Washingtonia robusta*) is particularly abundant. Scrap outer layers of tissue and twist. Leaves also make cups, mats, roofing, and thatching. C, N

Pampas grass, *Cortaderia* spp.: Pounding and stripping are required for these tough fibers. Used more for weaving than cordage. N

Paper: While newspaper is the most common paper to make cordage from, almost any type of paper will work.

Plastic: Plastic bags are the most common cordage material because of the ease of use, but any long flexible item will work.

Rush, *Juncus* spp.: Stripped or used dried and as it. N

Sagebrush, big, *Artemisia tridentata*: Outer and inner bark is crushed and scrunched. C, N

Silk tree or mimosa, *Albizia julibrissin*: Bast is extracted and retted. C, N

Sisal, *Agave sisalana*: Long leaves are scrapped and retted. Good for hats, mats, rope, twine, and weaving. C, N

Stinging nettle, *Urtica* spp.: Lightly pounded, stripped and used for light use. Good for thread and twine. The giant nettle, *Urtica dioica*, produces the most useful fiber. N

Figure 2.7   **Place mats and coasters woven from Mexican fan palm fronds Place mats and coasters were woven from Mexican fan palm fronds to enhance the experience of this outdoor meal.**

Figure 2.8   **This basket was woven from the young branches of the coppiced willow that grows to the right of it. Picture taken at the Marin Art and Garden Center, Ross, CA**

Sugarcane, *Saccharum officinarum*: Tough plant. Getting to the fibers includes crushing, shredding, and separating. Blades are better used for weaving. C, N

Ti, *Cordyline* spp.: Widely grown in urban areas. Freezing the leaves and stripping is common. C

Willow, *Salix* spp.: The bast is pulled from the outer bark for long fibers. The best bast come from shoots about ¼-inch in diameter; it is difficult to extract bast from old wood. N

Yucca, *Hesperoyucca whipplei* and *Yucca* spp.: Pounding, scraping and stripping the leaves are required for this strong fiber. Retting the scraped leaves creates a beautiful and pearly white strong fiber. C, N

Paper

While making your own paper can be energy and carbon efficient, and more so if you use discarded material such as household scraps and greenwaste, the tradeoff is time. Making paper is a lengthy endeavor.

Nomenclature

Bast: The fine fibers of a stalk or trunk (the phloem). Bast fibers produce prized paper.

Cellulose: The primary component of the cell walls of plants and of wood. It is a complex, but inert carbohydrate.

Couching: The act of removing paper from the mould and deckle.

Deckle: The frame that defines the shape and size of paper

Enhancements: Homemade paper is often enhanced with dyes, flowers, leaves and fragrances. Seeds can also be added to pulp to make a seed starter you can write on and then plant.

Fiber Preparation: Making pulp varies with the material. Using household paper involves shredding and blending. Harvesting plant fibers and removing its cellulose can involve drying, retting, chopping, beating, and simmering in an alkaline bath.

Mould: A fine mesh tray that scoops the pulp out of the vat.

Press: Whether using a mechanical press, passive press (like stacked cardboard), or ironing, pressing homemade paper reduces curling and can improve strength.

Pulp: A liquidity mass of long-stranded similar matter.

Vat: A container big enough to blend the fibers and hold the deckle and mould.

Window Screen: Often used to couch the paper from the deckle.

COMMON PLANTS FOR PAPER MAKING

While almost any plant can be used for paper making, those high in cellulose are the best. Notably, the easiest paper to make in urban areas is from paper itself: Cardboard, newspaper and wrapping paper make excellent everyday papers.

Legend
C: Cultivated
F: Found naturally

Bulrush and tule, *Bolboschoenus* spp. and *Schoenoplectus* spp. F

Cattail, *Typha* spp: Blades. C, F

Celery, *Apium graveolens*: Stalks. C

Crocosmia: Leaves. C, F

Fennel *Foeniculum vulgare*: Stems, stalks and leaves. C, F

Flax, *Linum* spp.: Stems, stalks and leaves. C

Grass, many varieties: Blades. C, F

Iris **spp**.: Leaves. C, F

Kitchen scraps: Plants high in cellulose. Celery, fennel, onion skins, etc. C

Mustard *Brassica* spp.: Stems, stalks and leaves. C, F

Palm fronds: Many varieties. C, F

Papyrus, *Cyperus papyrus*: C, F

Rush, *Juncus* spp.: Stalks. F

Sagebrush, *Artemisia tridentata*: Bast. C, F

Sentry plant, *Agave americana*: Outer skin of leaves. C

Stinging nettle, *Urtica* spp. Stem and stalks. C, F

Sunflower, *Helianthus* spp.: Inner stalk. C, F

Ti, *Cordyline* spp.: Leaves. C

Tree of heaven, *Ailanthus altissima*: Leaves and wood are high in cellulose. F

Religious and Seasonal Items

Having worked with students from across the world, I know that every culture has certain plants that are used for celebrations, ceremonies, holidays, and remembrances. To narrow down the hundreds of possible plants, I have chosen those that are widely used, can be easily grown, could offset the pollution from store-bought items, and might even be profitable to grow in urban areas.

PLANTS USED BY HOLIDAY

New Year's: Paperwhite (*Narcissus papyraceus*), wreathes (see Christmas below), and any indoor plant, such as peace lily (*Spathiphyllum wallisii*).

Figure 2.9 **Ceremonies are essential for community building, and the accessories that define them are just as vital. Pictured is an end-of-semester celebration at the Lyle Center for Regenerative Studies (Pomona, CA). Bouquets are an integral part of this ritual, and the Center actively grows them just for this purpose.**

Chinese New Year: Lucky bamboo (*Dracaena sanderiana*), ginger (*Zingiber officinale*), ginseng (*Panax* spp.), jade plant (*Crassula ovata*), money tree (*Pachira aquatica*), *Narcissus*, and orchid.

Valentines: Bromeliad, *Cyclamen*, orchid, rose (*Rosa* spp.), red anthurium (*Anthurium andraeanum*), and sweet violet (*Viola odorata*).

St. Patrick's Day: Clover (*Trifolium* spp.), *Iris*, naked ladies (*Amaryllis belladonna*), *Oxalis*, potato, and pot marigold (*Calendula officinalis*).

Passover and Easter: Cherry blossoms (*Prunus* spp.), daffodil (*Narcissus pseudonarcissus*), dogwood (*Cornus* spp.), hyacinth (*Hyacinthus orientalis*), *Iris*, lily (*Lilium* spp.), tulip, and all the dye plants for coloring eggs (see Dye section above).

Earth Day: Propagate and give away plants that produce a harvest or are native, or both.

Mother's Day: Baby's breath (*Gypsophila paniculate*), carnation (*Dianthus caryophyllus*), lilac (*Syringa vulgaris*), *Rosa*, and stock (*Matthiola incana*).

Independence Day: Baby's breath (*Gypsophila paniculate*), calla lily (*Zantedeschia aethiopica*), *Delphinium*, *Dianthus*, jasmine (*Jasminum officinale*), lavender (*Lavandula* spp.), melon, rosemary (*Rosmarinus officinalis*), *Rosa*, sea lavender/statice (*Limonium* spp.).

Yon Kipper: Grape, melon, sweet potato, resurrection plant (*Selaginella lepidophylla*), and tuberose (*Polianthes tuberosa*).

Halloween: Gourds, pumpkins and plants used in candies, such as horehound (*Marrubium vulgare*), licorice (*Glycyrrhiza glabra*), sugar maple (*Acer saccharum*), and sweet violet (*Viola odorata*).

Día de los Muertos: Orange marigold (*Tagetes* spp.), baby's breath (*Gypsophila paniculate*), cockscombs (*Celosia cristata*), corn (*Maize* spp.), and *Gladiolus*.

Thanksgiving: Corn, carrot, mint, persimmon, potato, pumpkin, walnut, yam, and grain crops. Large properties could grow poultry.

Hanukkah: Bay (*Umbellularia californica* and *Laurus nobilis*), date (*Phoenix dactylifera*), fig (*Ficus carica*), olive (*Olea europaea*), pomegranate (*Punica granatum*), yellow citron (*Citrus medica*), and grain crops, particularly wheat and barley.

Kwanzaa: Black-eyed pea or cowpea (*Vigna unguiculata*), collard greens or kale (*Brassica oleracea L. subsp. acephala*), corn, okra (*Abelmoschus esculentus*), yam (*Dioscorea cayenensis*), and mat making fiber crops (see previous Common Fiber Plants section).

Christmas: Of all the holidays, Christmas is the most consumptive. Below are some of the things that can be grown for celebration, gifts, and income.

Bouquets: Red *Amaryllis*, *Chrysanthemum*, Holly (*Ilex* spp.), mistletoe (*Phoradendron* spp.), paperwhite (*Narcissus papyraceus*), poinsettia (*Euphorbia pulcherrima*), *Rosa*, and corn husks, which are both decorative and used for tamales.

Wreathes: Bay (*Umbellularia californica* and *Laurus nobilis*), cypress (encompasses many genera), incense cedar (*Calocedrus decurrens*), juniper (*Juniperus* spp.), and western red cedar (*Thuja plicata*).

Christmas Trees: Many of the trees cut and used indoors during December can be grown in urban areas. The most common trees are:

- Leyland cypress (*x Cupressocyparis leylandii*).
- Spruce (*Picea* spp.): Colorado (blue), Norway and white.
- Pine (*Pinus* spp.): Aleppo, Monterey, Scotch and white.
- Fir (*Abies* spp. and *Pseudotsuga menziesii*): Balsam, Douglas, Fraser, noble, silver and white.

Figure 2.10 **This quarter acre farm grows Christmas trees along a busy freeway. While the prices are higher than industrially grown trees, community members appreciate watching their trees grow throughout the years. They have approximately 323 trees and harvest about 60 a year. Picture taken at Farmakis Farms, San Juan Capistrano, CA.**

Resources

BOOKS

Balls, Edward K. *Early Uses of California Plants.* University of California Press 1965.

Burgess, Rebecca. *Fibershed: Growing a Movement of Farmers, Fashion Activists, and Makers for a New Textile Economy.* Chelsea Green Publishing. 2019.

Campbell, Paul D. *Survival Skills of Native California.* Gibbs Smith. 2009.

Clarke, Charlotte Bringle. *Edible and Useful Plants of California.* University of California Press. 1977.

Funk, Alicia and Kaufman, Karin. *Living Wild: Gardening, Cooking and Healing with Native Plants of California.* Flicker Press. 2013.

Marilyn Wilkins. *California Dye Plants.* Thresh Publications. 1976.

Reid, Sara; Wishingrad, Van and McCabe, Stephen. *Plant Uses: California: Native American Uses of California Plants—Ethnobotany.* University of California Santa Cruz Arboretum. June 2009.

Sweet, Muriel. *Common Edible and Useful Plants of the West.* Naturegraph Publishers, Inc. 1976.

Timbrook, Jan. *Chumash Ethnobotany: Plant Knowledge among the Chumash people of Southern California.* Heyday Books 2007.

WEBSITES

"Desert Harvesters," a non-profit, grassroots effort at www.desertharvesters.org
"Fibershed," a nonprofit grassroots effort at www.fibershed.org
"Plants for a Future," a non-profit at www.pfaf.org

Note

1 This chapter was written with the help of the resources below, along with a lot of time, trial, error and observation. Much of this chapter was cooked up in the kitchen at the Lyle Center for Regenerative Studies at California Polytechnic University, Pomona.

Chapter 3

Energy

Fossil fuels are notoriously inefficient. Each gallon of gasoline we draw from the earth required about 90 tons of prehistoric algae, plants, and other forms of organic life. To put this into perspective, 90 tons of vegetative matter today has the energy equivalent of 5,100 gallons of gasoline.[1]

But it's not the inefficiencies of fossils that are dragging us down; it's their nature. Fossils are prehistoric in origin. Continuing to rely on them takes us backward; relying on renewable energy moves us forward. Why use a product that is extinguished once you use it when you can use something that is circular? This chapter explores that forward-looking path—local renewable energies.

Harvesting local energies is the path forward. However, generating renewable energies in urban areas is much different than in rural or agricultural areas. Land has greater value and more uses in urban areas. Labor costs are generally higher too. And the scale at which energy capture occurs is also different. Industrial energy will devote thousands of acres to a project, whereas a municipal project may only have a few hundred.

Covered below are the most common types of energy captured, grown and employed, which includes biofuels, earth, human, solar, and wind energies. Also included is a brief overview of storing electricity without the use of chemical/rare Earth batteries and a discussion of ERoEIs.

Strategies

Before starting any energy-capturing endeavor, consider these three strategies.

Prioritize Passive Energies: Passive energies are those that do not require extensive processing, infrastructure and/or materials. Examples would include using thermal mass to regulate a building's temperature, allowing the sun to light a structure, and funneling wind to cool a building.

DOI: 10.4324/9781003369752-3

Figure 3.1   **This small and dark office does not need industrial energy to be functional during the day. It is brightened by a solar tube. These daylighting devices are useful even in winter and on cloudy days. Simple, long lived, and effective define passive energies.**

Run an ERoEI: ERoEI stands for "energy returned on energy invested" and is a measure used across the energy industry. The goal is that the energy produced is greater than the energy invested to capture it. Whether running a bulldozer to move earth or installing PV panels, make sure that the site, technology, and levels of maintenance can facilitate no less than adequate ERoEIs. Refer to the end of this chapter for greater detail of ERoEIs.

Reimagine Conservation: Deprivation is not the goal of this book. To enjoy all the elements of the good life—health, happiness, and prosperity—we must consume. That's a given. But we need to consume intelligently. We need to maximize passive forms of energy and all forms of local renewable energy. We also need to use every bit of energy to its greatest benefit. Consumption, like conversation, is an art.

Nomenclature

AC: This means alternating current. It is an electric energy wave that can change directions (polarity reverses), unlike DC. AC is high-quality and travels well. Wind turbines produce AC.

Aerobic Respiration: The decomposition of organic material by bacteria or fungi in the presence of oxygen.

Alternator: A device that turns mechanical (kinetic) energy into electricity. They produce AC and high-quality electricity. They cannot charge batteries.

Anaerobic Respiration: The decomposition of organic material by bacteria or fungi in the absence of oxygen. Methane is a product of anaerobic respiration.

Biofuels: Any energy source derived from biomass, which can be either plant or animal.

Biomass: Technically, biomass is organic matter that has been completely dried. Commonly, biomass is anything organic, whether wet or dry, that can be utilized as a fuel.

British Thermal Unit (BTU): A measure used to express energy content. A BTU has the energy equivalent of burning a wooden match to ash or raising the temperature of 1 pound of water 1°F at sea level.

Calorie: See Kilocalorie.

Cellular Respiration: The breakdown of chemical energy, such as organic material, which results in the release of a different form of chemical energy.

Chemical Energy: The potential energy embedded in the chemical bonds or nature of certain materials, such as wood and hydrogen (H_2).

Convective Energy: The movement of energy (heat) through air or water. Warm air or water rises, cold air or water falls, and both create a potential energy that can be utilized.

Cord of Wood: One cord is 128 cubic feet of wood, which is a pile about 4 feet wide, 8 feet long, and 4 feet deep. Firewood and sometimes pulpwood are bought and sold using the Cord unit volume. There are 3 cords in 1,000 board feet of hardwood and 2.8 cords in 1,000 board feet of pine.

Cubic Foot of Natural Gas: A unit of energy that is used to express the sale of natural gas. One cubic foot has the energy equivalent of about 1,031 BTUs.

DC: This means direct current. It is an electric energy wave that moves in the same direction (polarity never reverses), unlike AC which moves in both. DC provides low quality energy. PV panels produce DC.

ERoEI: An equation that measures the efficiency of an energy source. It means energy returned on energy invested.

Feedstock: The organic material grown or gathered that becomes an energy source for something, such as biofuels, grazing animals, and microscopic life.

Generator: A device that turns mechanical (kinetic) energy into electricity. They can produce either AC or DC, but the energy is low quality. Generators that produce DC electricity are used to charge batteries.

High Quality Energy: A powerful source of energy. Gasoline is considered high quality because it has 125,000 BTUs per gallon. Other high content/density energies include AC electricity, biogas, and ethanol.

Joule (J): A unit of energy equal to .000947 BTUs. Joule is used to express electricity, heat and work.

Kilocalorie (kcal): A unit of energy used to measure the energy content of food. Commonly referred to as a Calorie. One kilocalorie is equal to 3.97 BTUs. Humans need approximately 2,000 to 3,000 calories (kcal) a day.

Kilowatt: A measure of electrical energy that is equivalent to 1,000 watts and 3,412.14 BTUs.

Kilowatt Hour (kWh): A measure of electrical energy. It is equal to using a 100-watt lightbulb for one hour and has the energy equivalent of 3,412.12 BTUs.

Kinetic Energy: The energy in motion. Human power, flowing water, and wind all have kinetic energy. Kinetic energy can be converted to electricity and/or work.

Low Quality Energy: A low intensity form of energy. Low quality energy is capable of low-quality work. Low content/density energies include compost heat, daylighting, and thermal mass.

Mechanical Energy: An energy derived from motion. It is also called kinetic energy. For instance, burning wood (chemical energy) boils water that produces steam that drives a turbine (kinetic energy), which produces electricity.

Megawatt (MW): A measure of electrical energy and the same as 1,000 kilowatts or 1,000,000 watts. One megawatt is enough to power between 400 and 1,000 average homes in the U.S. (an accepted standard is 750 homes) The large variance is due to the differences in climate and their subsequent impacts on energy use.[2]

Methane: CH_4 is an odorless, colorless, and flammable gas. It is the result of organic matter being decomposed in the absence of oxygen. Often referred to as Biogas.

Passive Energies: Energies that do not require extensive processing, infrastructure and/or materials. They are typically low-quality.

Radiant Energy: Energy traveling in electromagnetic waves. The most common expressions of radiant energy are heat and light.

Solar Energy: Radiant energy produced by the sun. Solar (radiant) energy can be converted to electricity, heat, and photosynthesis (chemical).

Therm: A unit of measure for natural gas. One therm is equal to roughly 97 cubic feet of gas and 100,000 BTUs.

Thermal Mass: The density of a material that helps stabilize fluctuating temperatures. The greater the mass, the greater the stabilization. Earth, concrete blocks, and water are some of the materials with high thermal mass.

Watt: A measure of electrical energy that is equivalent to 3.412 BTUs and 1 Joule per second.

Wind Energy: Air currents are a kinetic energy that can be utilized in a variety of ways. Wind energy (kinetic) can be converted to cooling, electricity, and work.

Urban Energy

Listed below are the best sources of renewable energy in urban areas, which include biomass, earth, solar, and wind.

Biomass

Biofuels are made from biomass that is harvested or gathered from current living things, not from things that lived in the past. Biofuels can be made from plant and animal parts, and there are many varieties.

PASSIVE BIOFUELS

These are the fuels that require minimum processing. The two most prevalent are compost and firewood.

Compost

Uses: Compost is used as a heating energy. Compost temperatures range from 90°F for small piles to 140°F and more for large and active piles. Compost is used to heat air, water, or piled against a structure. All three uses of compost can raise indoor temperatures by 1°F to 5°F.

Costs: Low to moderate.

Difficulty: Not difficult to understand. Expertise that might be needed includes HVAC, plumbing and/or masonry.

Best Scale: Large-scale. Generating a significant amount of transferable heat requires thousands of pounds of compost. Only large properties can generate that much material. The ideal site has an enormous amount of organic waste and small structures that need warming.

How-To: There are three primary methods of capturing and distributing compost heat.[3]

1. An air-based method of heat capture pulls air through the heart of a compost pile where it is then pushed through ducts, typically exchanging its heat with water. These systems are the least common. System components include large blowers, ducts, and a heat exchanger.
2. A water-based method of heat capture involves layering metal pipes in a pile and pumping the heated water to its designation, which might be a radiant flooring system, a water heater, radiator (space heater), or fluid-to-air radiator. In addition to the large amount of plumbing, these systems also require a pump and surge tank. This type of system can increase the temperatures of very small structures, such as a hoop house, by 3°F.[4]

3. Compost is piled up against a structure to warm it. Depending on a multitude of factors, such as the size of pile, size of structure and the environment immediately around the pile, compost has the potential to raise indoor temperatures 1°F to 5°F. Piled compost also shields a building from cold wind and provides some thermal mass, both of which reduce heat loss from the structure.

Notes: A long lasting heat generating pile will have the proper ratio of carbon to nitrogen (2–3:1), material that will readily decompose (no tree trunks), appropriate moisture content (40% to 60%), and placement and maintenance that will protect it from rain, snow, and extreme cold and heat.

Firewood

Uses: Used for cooking, heat, light.

Costs: Low.

Difficulty: Not difficult. We've been doing it for eons.

Best Scale: Small scale. Burning wood for heat is best at the residential scale.

How-To: A 1,000 square foot structure needs about 3 cords of wood to warm it in cold winter climates; mild-winter climates need much less. A cord is 128 cubic feet of wood, or about half of what can fit in a small pickup.[5] Hardwoods burn hotter and longer than softwoods but take longer to cure and dry. Refer to the Timber chapter for more detail.

Notes: Using wood for heat can be sustainable, carbon neutral, and might be a widely available resource. However, burning wood is bad for air quality because of the particulates and gaseous pollution. Notably, it is illegal to burn wood on "Spare the Air Alert Days". For information on how to cleanly burn wood visit EPA's "Burn Wise" website at https://www.epa.gov/burnwise.

Figure 3.2 **Burning wood for cooking, heating, and lighting is ancestral. Humans are deeply connected to an open flame. Moreover, the carbon released is new, not ancient, like industrial energies. Pictured is a pizza oven, a fantastic contraption for community building.**

FOOD AND HUMAN ENERGY

At best humans turn 20% of the food they eat into work and the power they exert is small. Yet despite the low output, humans are 90+% more efficient than fossil fuel driven work. However, the efficiency of human labor is directly tied to the energy costs of food. Energy drenched foods, such as beef and processed foods, reduces the efficiencies of human labor. The discussion below examines the potential of human labor, not the energy costs of food production. Please, refer to the Food chapter.

HUMAN ENERGY CAN BE EMPLOYED THREE WAYS

- *Direct Energy*: Humans use their energy to supplant fossil-fueled machines, which includes digging, harvesting, mowing, pruning, sickling, sweeping, tilling, weeding, and much more.
- *Kinetic Energy to Kinetic Energy*: Human energy is used to run a device that runs a process. The most common devices are a bicycle, foot or hand crank connected to a process, such as a drill, grain huller, grinder, potter's wheel, pump (air and water), saw, sewing machine, thresher and/or winch.
- *Kinetic Energy to Electricity*: Human energy is turned into electricity via a crank and generator. Although common, it is not an efficient use of human labor. Depending on the machine, people will lose between 42% and 67% of their energy output. The alternator/generator, regulator, and wires create a great deal of loss, more so if a battery and/or converter (turning AC to DC) is attached to the system. Bicycles and hand cranks are the most common ways to transfer human energy to electricity.

HUMAN ENERGY TO ELECTRICITY

Turning human energy into electricity is not efficient. Pedaling a bike at a moderate pace for an hour can produce 100 watts of power. If this energy were linked to an alternator, here are some of the things humans can power:

AC Energy

- 1-80w light bulb or 2-40w light bulbs and 2 rooms.
- DC Energy[6]
- 1-router at 10W.
- 1-laptop between 20W and 50W.
- 1-smartphone at 2.5W and 5W.
- And 5–6-watt LED light bulbs.

Table 3.1   Human Daily Requirements

Required Daily	Female	Male
Calories (kCal)	2,000	2,500
BTUs	7,931	9,914
Biomass in pounds	.92	1.15
Daily Gas Exchange in Pounds	.65	.81
O$_2$ pulled down	.83	1.04
CO$_2$ released		
Output: Possible Work	399.5	499.5
Calories per 8 hours	1,586/465	1,983/581
BTUs/Watts per 8 hours	58	73
Watts per Hour		

All the figures above are industry standards and a broad average.[7] To help, 1 pound of biomass represents 2,166 calories and 8,600 BTUs.

HUMAN TRANSPORTATION COMPARED

The graph below showcases the differences between human and fossil fuel driven transportation. This graph also explains why human energy has been supplanted by fossil fuels, despite the huge energy and environmental costs—human labor is far more expensive than fossil fuels.

Table 3.2   Human Transportation Compared

Transportation: Travelling 5 Miles	Energy Used: Gasoline, calories, BTUs	Carbon Emitted	Time to Accomplish	Cost of Fuel and Labor
Drive: 24mpg	.21 gallons 6,560 calories 26,043 BTUs	4.2 pounds	7.5 minutes at 40mph	Fuel: $0.84 Labor: $2.67
Walk	.012 gallons 366.5 calories 1,454.9 BTUs	.24 pounds	86 minutes at 3.5mph	Food: $1.52 Labor: $29.20
Bike	.0054 gallons 170.8 calories 678 BTUs	.11	25 minutes at 12mph	Food: $0.43 Labor: $8.33

The figures below will help make sense of the graph above. These figures are standard averages.[8]
I gallon of gas represents 31,500 calories (kcal) and 125,000 BTUs.
Consuming I gallon of gas emits 20 pounds of CO$_2$.
Walking uses 255.7 calories an hour and biking 409.9 per hour.
Gasoline is $4 a gallon, food is $25 a day ($1.04 per hour), and labor is $20 an hour.

ACTIVE BIOFUELS

These are the biofuels that require significant processing. The four common types are biodiesel, biogas, biomass for electricity, and fermentation.

Figure 3.3   **Bicyclists average about 10 mph, almost the same as city buses. When compared to any transport powered by industrial energy, bicycles are low impact and incredibly energy efficient. Picture taken at a CicLAvia event in downtown Los Angeles.**

Biodiesel

Uses: Biodiesel is turned into electricity, heat, light, transportation.

Costs: High.

Difficulty: Requires training and specialized equipment.

Best Scale: Small to large scale. Producing biodiesel from a waste product, such as vegetable or animal oils from food processing, makes sense at any scale (see caveat below). Growing the feedstock for biodiesel, like soybean or sunflower, does not make sense in urban areas and is only for large scale.

How-To: Biodiesel is made from vegetable oil and animal fats, whether these oils are cultivated or gathered as waste. Plants commonly used for biodiesel, and those that can be grown in urban areas, include algae, castor bean, flax, hemp, jatropha, jojoba, mustard, pennycress, pepperweed, radish (wild), rapeseed, soybean, and sunflower. Vegetable oil waste and the fats from food processing are a more common source of feedstock in urban areas because they are available in large quantities and are easier to process.

Pros

- Biodiesel produces less CO_2 per mile than petroleum diesel.[9]
- Making biodiesel from waste products closes a loop and makes ecological and economic sense.
- Before biodiesel became strictly regulated, especially in California, biodiesel was the people's energy; anybody could refine it if they had a source for feedstock and a modified engine.
- Biodiesel is less damaging to the planet than ethanol.[10]

Cons

- Homemade biodiesel has been outlawed in California. You can buy biodiesel, but it has met strict industry specifications. Burning raw vegetable oil is illegal in the U.S. All diesels (petroleum and animal/plant) are high in nitrogen oxides, which are dangerous in their first stage, but when these gases react to sunlight and heat, they turn into ozone and cause many types of respiratory problems.
- Besides requiring hundreds of acres of cropland to make economic sense, the process of production is lengthy: oils need to be separated from the feedstock, glycerin needs to be separated from the oils, and then sulfur oxides need to be scrubbed from the fuel. As a waste product biodiesel has good ERoEIs, but as a grown product they do not have good ERoEIs.
- Growing the feedstock for biodiesel can increase the number of croplands, and cropland is the number one destroyer of native plants and animals. Growing feedstock might also increase the price of food and other cropland commodities.

Biogas

Uses: Biogas is used for cooking, electricity, heat, light, transportation.

Costs: Moderate to high

Difficulty: Producing methane is easy; capturing it for practical use is not.

Best Scale: Large scale. Biogas generation and use is best at large scales, such as facilities that handle industrial and agricultural waste. While feasible, residential systems need access to the most digestible material, which includes sewage, food waste, and only the greenest and thinnest greenwaste, such as grass clippings, leaves and twigs. Few small properties are equipped to handle sewage.

How-To: Biogas happens naturally. It is the product of organic matter being broken down by bacteria without oxygen (anaerobic cellular respiration). Biogas, which is methane, can be made from almost anything organic. Urban greenwaste, animal farm byproducts (body parts and manure), food waste, sewage, and almost any type of crop residue can be feedstock. However, the more digestible the feedstock, the better the economics. Woody feedstock, such as urban greenwaste, is too slow to break down to make economic sense.[11]

As a production fuel biogas is made in a biodigester. These devices can be simple or complex, large or small. A simple device will produce gas with about 600 BTUs per cubic yard. With refinement a complicated system can achieve 1,000 BTUs per cubic yard, which is equivalent to natural gas. One gallon of gasoline has the energy equivalent of 208 pounds of cow manure digested into biogas.[12]

The best efficiencies for biogas production in urban areas are where the feedstock is easily digestible, abundant, and the biogas can be used onsite to power a need. Large cow and poultry farms have some of the most successful systems; bedding, feed waste, and manure are funneled into a generation system to warm structures and water. A good example is the Los Angeles County Sanitation District's Joint Water Pollution Control Plant that produces electricity, heat/steam, and vehicle fuel from 30 tons of commercial food waste a day.

Biomass to Electricity

Uses: Burning biomass to generate electricity.

Costs: High.

Difficulty: Although steam turbines have been around for hundreds of years and the process is easy to understand, it requires engineering skills.

Best Scale: Small and large scale. Any site that has an ample and steady supply of woody material is ideal. Small systems run at about 1 ton of biomass per hour, medium systems 10 tons per hour, and large 500 tons per hour. Roughly 1 ton of dry biomass produces 1 megawatt.

Cities can employ these technologies, but not necessarily as an efficient means of energy generation, but for waste disposal. Cities can use portable devices that consume 1 to 10 tons of biomass an hour and produce 100kw to 1,000kw, which is not efficient, but when the reductions in handling and transportation of the material are added, the benefits often outweigh the economic and energetic costs.

How-To: Biomass to electricity is done in one of two ways: direct combustion or gasification. Direct combustion has been around for hundreds of years and involves burning wood to boil water that produces steam that turns a turbine and generates electricity (15% to 20% efficiency). Gasification involves super heating biomass in an oxygen-starved environment and capturing the energy rich biogas for fuel, which is then used for cooking, electricity, and transportation (7% to 14% efficiency[13]). The efficiencies of both methods rise greatly if the heat generated from the process is employed elsewhere onsite, called cogeneration.

The feedstock for biomass energy plants is diverse and can include agriculture waste, construction and demolition debris, forest waste, land and road clearing debris, mills (mostly lumber and paper), urban greenwaste, and/or processed material, such as wood pellets.[14]

Pros

- Constant electricity generation, unlike solar or wind.
- The feedstock can be all waste, reducing hauling and landfilling costs.

- Can support a circular economy and carbon cycle.
- Wood fuel contains low amounts of sulfur and heavy metals, which reduces acid rain when compared to fossil fuels.

Cons

- Any type of incineration contributes to air quality problems and climate change.
- The waste products—ash, condensates, oil and water—contain toxins.
- Electricity from biomass is more expensive than fossil fuels and electricity from solar and wind.
- Because of the low energy density of wood (compared to fossil fuels) electricity generation demands huge storage areas, which increases handling costs and the footprint of the facility.
- Maintenance costs are high because of feedstock management and ash removal.

Ways to increase efficiency:[15]

- Cogeneration: An energy producing plant should use the heat it produces from energy generation for other things, such as heating buildings, heating water, and/or drying feedstock.
- The electricity and heat can be used onsite.

Figure 3.4 **FireBoxes (Air Burner, Inc.) are generating electricity from the snags left over from the Dixie Fire in Northern California, 2021. Because of the incredible scale of destruction, generating electricity and nutrient rich ash from the snags made better sense than trying to mulch and/or compost the miles of burnt trees.**

- The feedstock is abundant year around and within 50 miles of the facility.
- High quality feedstock. The efficiency of a system hinges on the quality of feedstock, and quality hinges on dryness and density. The drier and denser the wood going in, the more electricity that flows out.

Fermentation

Uses: Ethanol and methanol are used for cooking, electricity, heat, light, transportation.

Costs: High

Difficulty: While any sugary vegetation can be fermented, the process is complicated and requires specialized equipment.

Best Scale: Large scale. Producing fermented energy (ethanol or methanol) to meet urban needs is best at large scales and on large farms. Urban waste does not produce the feedstock for high-quality fuel. Most fermented energy comes from feedstock that was grown.

How-To: Fermentation produces a combustible liquid and occurs when bacteria or yeast breaks down sugars. The process is anaerobic, like biogas. Yeast (fungi) produces drinkable alcohols and ethanol. Bacteria produces methanol. Crops used for feedstock are high in sugars and include algae, corn, cattails, euphorbia, grapes, grains (like wheat), millet, sorghum, sugar beets, sugarcane, and large grasses, like *Miscanthus* and switchgrass.

Ethanol and methanol are excellent fuels because of transportability, combustibility, and the fact that feedstock can be grown. However, fermentation is not a fuel that can be easily produced in urban areas.

Notes: Fermentation is an excellent idea in cities, but not as a fuel source. Urban folks should be fermenting and preserving foods, creating craft liquors, health foods, and vinegar. Urbanites have the money, means, and motivation for fermented innovations. The energy content of ethanol and methanol are fantastic, but the process demands feedstock that cities struggle to produce.

Earth

Our mother is a provider, and she has long given comfort and energy in her mantle. Whether helping indigenous peoples stay warm and cook or providing energy to millions of modern people, the land below our feet has much to offer.

PASSIVE EARTH ENERGIES

Thermal Mass

Uses: Thermal mass provides comfort in both cold and hot environments. Earth provides these benefits through insulation (protecting and shielding) and immense volume (absorbing).

Figure 3.5   **An illustrated summary of the types of earth energies that can be utilized in urban areas. The three types covered below include thermal mass for thermal comfort, geothermal for temperature regulation, and geothermal converted to electricity.**

Costs: Low to high.

Difficulty: The theory is easy and moving soil is not difficult but building structures that can withstand the pressures of soil, roots and water requires an architect/engineer.

Best Scale: Small and large scale. From root cellars to large commercial complexes, using the earth to regulate temperatures has been used for centuries. Earth regulation is applicable at every scale: residential, commercial and municipal.

How-To: Because the goal is to root a process, such as a structure, into the Earth, there are only two techniques: digging down or building up.[16]

Digging Down

Sinking a structure into the ground is fantastic for temperature regulation. The temperature of soil 5 feet down typically ranges between 55°F and 72°F in California (low deserts and areas above 6,000 feet deviate from this range).[17] Cooling or heating a structure when two of the walls are a constant temperature is far easier than heating four walks exposed to atmospheric variation.

However, there are two drawbacks to sinking structures. First, excavation and reinforcing a structure against the earth and water are expensive. Second, unless properly constructed, a structure may be damper and consequently have mildew and mold, which is neither good for the health of the occupants nor structure.

Building Up

Mounding soil against a structure to regulate temperatures is easy and requires less structural reinforcement than going down. Soil in this manner should be thought of as a

Figure 3.6   **Building up and insulating this building with soil provides many benefits. It cools the warm structure and entryway. It provides more space for planting. And it is aesthetically interesting and unique. Picture taken at Roger's Gardens, Corona Del Mar, CA.**

blanket and the thicker the blanket, the more effective at moderating temperatures. The wall retaining the soil will need reinforcement and protection from moisture. Built up soils also dry quicker than natural soils and will require more irrigation if planted.

ACTIVE EARTH ENERGIES

Pipes and Tubes for Geothermal Cooling and Heating

Uses: Buildings are cooled or heated using the constant temperature of the Earth's crust.

Costs: High upfront costs but low operating costs.

Difficulty: The theory is easy to grasp but designing and plumbing the system is difficult. Every system must be tailored to the land and buildings.

Best Scale: Small and large scale. Because of the high initial costs these systems are best where the owners plan to occupy the structure for 15 or more years and can recover their investment. Good candidates include commercial complexes, government buildings, and schools. Single story buildings may find thermal massing (discussed above) more cost effective.

How-To: The goal is to exchange the cold or heat in a building with the constant 50°F to 60°F deep underground. The two primary ways to facilitate this exchange are pipes or tubes. Pipes are used for water and tubes for air. These pipes or tubes are either buried horizontally or vertically. If horizontal, trenches range from 8 feet to 30 feet deep; if vertical, wells are 100 feet to 500 feet deep. The more plumbing underground, the better the heat exchange.[18] These types of systems are often used in partnership with conventional cooling and heating systems.

Note: Geothermal HVAC systems are generally 25% to 50% more efficient than conventional air-source systems.

Geothermal Heat for Electricity

Uses: The heat and/or steam linked to earth's magma are plumbed to either feed heat directly to a process or to drive a turbine for electricity.

Costs: High.

Difficulty: The technology is simple but finding a location where geothermal heat is near the surface and not already claimed is difficult.

Best Scale: Large scale. Geothermal is a local energy, but not necessarily an urban energy. Many communities get their electricity from geothermal, but it would be difficult for a city, or a commercial and residential property to develop such a resource. With that said, the technology scales down well and small plants have proven cost effective.[19]

How-To: Geothermal plants drill 1 or 2 miles deep and install two pipes, one for extraction and the other for injection. In the extraction pipe hot steam or water flows upwards and drives a process, such as a turbine. In the other pipe the condensed water is pumped back into the earth. On a small scale, there are many places where people can jump into a hot spring heated by geothermal energy.

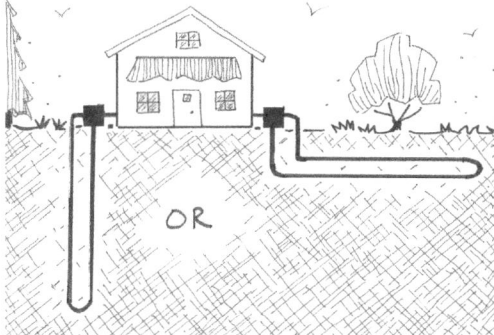

Figure 3.7 **Illustrated are the primary types of geothermal heat exchange: vertical and horizonal. The underground pipes are filled with air, fluids, or refrigerants.**

Solar

Except for Earth energies, the sun drives every other energy in this chapter. PV panels and thermal converters, biomass and wind, animal and human energy—all are powered by the sun. Below are the types of energies that can harvested from solar radiation.

PASSIVE SOLAR ENERGIES

These are solar energies that require minimum processing and the four most common are daylighting, drying, ovens, and thermal converters (hot water heaters).

Daylighting

Uses: Lighting the interior of a building with sunlight instead of light bulbs.

Costs: Low to moderate.

Difficulty: Not difficult to understand, but the design and installation can be tricky.

Best Scale: Small and large scales. Any place that uses light bulbs in the daytime is a candidate for daylighting. This is an easy technology to either design or retrofit.

How-To: Daylighting a building involves harvesting direct sunlight, diffused sunlight and sunlight reflected from an external surface. The largest drawback with daylighting is its impact on interior temperatures; improperly designed daylighting will create greater fluctuations in interior temperatures, leading to higher cooling and heating costs. Some of the common devices that balance lighting with heat gain include:[20]

- *Light-Guiding Shades*: Blocking the sun from directly striking an interior, light-guiding shades are exterior features that redirect sunlight to a ceiling, causing it to bounce throughout a room. This is a practical technology for hot environments, like the Southwest.

Figure 3.8   **Illustrated above are the types of solar energy that can be utilized in urban areas. The solar energies showcased below include daylighting (illustrated as a solar tube), drying, thermal converters for cooking and heating, concentrated solar, and photovoltaic (PV) panels.**

- *Light Shelve*: A device that allows the winter sunlight to penetrate a structure, but redirects the hotter summer sunlight to the ceiling, diffusing and cooling the light. A light-shelf can be either exterior or interior.
- *Prismatic Panels*: Taking direct sunlight and diffusing and redirecting it into the farther reaches of a room are what make prism panels effective.
- *Reflecting Wall*: Used to light up the north sides of buildings, reflecting walls are built far enough away from a structure to bounce the low winter light back into a structure. In a city a reflecting wall may be your neighbor's building.
- *Solar Tubes*: Called by many names solar tubes are cylinders measuring 10-inch to 2.5-foot in diameter that directs sunlight from a roof through a structure. These tubes can be up to 100 feet long. They are great for lighting small spaces, such as closets, hallways, and restrooms. Because of their small footprint, these devices have a low impact on indoor temperatures.
- *Skylight*: Providing either direct or diffused sunlight, skylights are centuries old and are ideal for single story buildings.

Notes: Daylighting is not just about energy conservation—it's about our health. Sunlight is essential for our wellbeing. Sturdier bones, improved morale, reinforced immunity, better sleep, and even lower incidences of some types of cancer are many of the benefits of welcoming sunshine indoors.[21]

Figure 3.9 **Regenerative structures are built to work with the landscape and increase the amount of natural light inside the structure. The daylighting strategies showcased above include many transparent surfaces (windows), shading devices that do not block the winter sun, skylight, solar tube, and reflecting wall.**

Drying

Uses: Dry foods, herbs and textiles.

Costs: Low.

Difficulty: Not difficult to understand or construct.

Best Scale: Small scale. Residential properties are the best candidates for drying because they have the space and time. That said, commercial enterprises in areas where there is more space might be able to pursue air-drying.

How-To: Drying anything involves warmth and air circulation. It does not depend on direct sunlight, which means that drying is as ideal indoors as outdoors.

Fabrics: Hanging fabrics to dry outdoors takes between 1 and 3 hours. Drying indoors requires 2 to 4 hours. Notably, a fabric's colors and fibers last longer when air dried, as opposed to either machine or sun dried.

Food and Herbs: The two types of drying are air and sun, and their approaches differ:

- Air-drying typically occurs indoors and requires warmth and mild air circulation. While drying herbs indoors is common, drying food is not because most spoils. Along with herbs, hot peppers and mushrooms are commonly air-dried.
- Sun-drying requires humidity below 60%, summer temperatures, direct sunlight, and good air circulation. With a few exceptions, most food will need cleaning and some type of preparation before being put in the sun. Fruits, greens, meats, peppers and tomatoes are some of the best candidates for sun drying.

Solar Ovens

Uses: Bake and cook food.

Costs: Low to moderate.

Difficulty: Not difficult to understand or construct.

Best Scale: Small scale. Solar ovens are fantastic on a small scale and in sunny environments.

How-To: Solar ovens are a low-tech concentrating solar panel (see further on). They are created from many types of configurations, such as box, dish, cylindrical or parabolic. Ideally, the food is baked or cooked in an enclosed glass box or cylinder, protected from the cooling effects of airflow.

Notes: Solar ovens have many advantages: They have no fuel costs and provide nearly carbon-free cooking (minus embodied costs); they travel well and set up fast; and they can be designed and constructed by the average person using salvaged materials. But they also have some disadvantages: they only cook when the sun is out; they struggle with dense food in cloudy and low elevation sunlight; and they are capable of fluctuating temperatures on cloudy/windy days because they lack the ability to hold heat (thermal mass).

Thermal Converters

Uses: Heat water and warm structures.

Costs: Low to moderate.

Difficulty: They are neither difficult to understand nor construct.

Best Scale: Small and large scale. While thermal converters work at any scale, they are best where the demand for hot water is high, such as apartments and dormitories, industrial processes and food processing. Heating water in U.S. homes represents about 19% of its total energy use.[22]

How-To: Thermal converters are typically non-concentrated, which means they are pipes enclosed in a plate installed on a roof. Pipes are filled with pumped air or fluid. These systems can be closed (recirculating) or open (replenishment of air or water). Concentrated solar, using a parabolic device to focus more of the sun's rays on the pipes, is the most effective, but also the most expensive and difficult to construct.

Figure 3.10   **This remote workstation relies on a thermal converter for its hot water. The system can be maintained and repaired with limited experience and skills, which further enhances its reliability. Picture taken at Humboldt Bay National Wildlife Refuge, Loleta, CA.**

Notes: Whether heating water for indoor use, pools, or melting metals, focusing the sun's energy on one task can deliver incredible results. The U.S. Department of Energy says that thermal converters can reduce the costs of heating water by 50% to 80%, which means they can pay for themselves in 5 to 10 years. Naturally, the sunnier the area, the better the ERoEIs.

ACTIVE SOLAR ENERGIES

These are solar energies that require manufacturing, large energy imports, and/or rare and/or toxic materials. The two most common in urban areas are concentrated solar and photovoltaic panels.

Concentrating Solar Power (CSP)

Uses: Produce electricity.

Costs: High.

Difficulty: The concept is easy to understand, but the mechanics of design and construction are difficult.

Best Scale: Largescale. The larger the project, the better the economies of scale and ERoEIs. Small-scale systems can be economically feasible, but the energy receiver should be local and there needs to be cogeneration (using both the electricity and heat). It also helps if the need is rural, or off the grid, and not competing with industrial sources of energy.

How-To: Concentrating solar power works by reflecting sunlight on one specific point, which is generally a tube filled with heat-transferring fluids, such as water, molten salt, molten silicon or synthetic oil. In some cases, the difference between the reflecting surface and the absorbing surface can be 100 times, creating temperatures of more than 750°F. CSP uses this heat to produce steam, drive a turbine, and generate electricity. A big drawback with CSP is the substantial water requirement:[23]

Photovoltaic (PV)

Uses: Electricity.

Costs: High.

Difficulty: Neither the manufacturing, installation nor repair can be handled by the typical person. PV electricity is industrial energy.

Best Scale: Small to large scale. From a price point of view, PV panels can be cost competitive throughout urban areas, whether commercial, industrial or residential. They can be integrated into the urban fabric.[24]

How-To: First, calculate your electricity need. Second, input your need and zip code in one of the many solar calculators to get a rough estimate of the number of panels and costs. PV panels in sunny areas, like the Southwest, produce between 14 and 18 watts of electricity per square foot.

Drawbacks

There are many disadvantages with photovoltaic panels: [25]

- One solar panel cannot make another solar panel, which means that they are not energy breeders. In fact, not one manufacturer is making PV panels using the energy from PV panels (although a portion of the industrial energy used may be comprised of solar power).
- They are constructed from finite and rare earth materials.
- They are toxic to mine, manufacture, and dispose.
- They demand a lot of space.
- And they cannot be built and maintained by the average person.

The use of PV panels should be minimized because of their many disadvantages. The first strategy for any solar endeavor is to maximize passive strategies—employ fewer toxic technologies. Passive strategies can cool and heat a building, provide natural lighting, dry fabrics, cook food, and heat water.

Best Uses

The best use for PV panels in urban areas is when they can do more than just generate electricity.

- **Cooling**: Use for cooling and shade in areas such as parking lots, carports and courtyards.

Figure 3.11   **Photovoltaic panels can be utilized for more than just electricity. Parking lots are a great example. They shade a notoriously hot area. They help slow the aging of asphalt and cars. And they shield fluid leaks, such as oil and radiator fluid, from rainfall and help prevent those toxins from flowing into the storm drain system.**

- **Rainwater Collection**: Use as a collection surface for rainwater. As an example, 100 square feet can direct 55 gallons of rainwater in 1 inch of rain: 660 gallons with 12 inches of rain.
- **Resiliency**: Use in areas not connected to the regional grid, which might be lone structures and safety features, such as signs and streetlights.
- **Alternative**: Use where the fuel source for electricity is toxic, as is the case with diesel and kerosene.
- **No Other Use**: Use on land that is severely degraded and/or toxic (brownfields) and could not support other active uses, such as food production and recreation.

Wind Energies

Plants have employed wind for millions of years, humans for tens of thousands. It is an ancient energy. Many plants use wind for pollination and/or seed dispersal. Humans can employ it urban areas to cool, dry, promote public health, transport, supply mechanical processes, and generate electricity.

PASSIVE WIND ENERGIES

These are wind energies that require minimal materials and manufacturing and include cooling, public health and transportation.

Cooling

Uses: Cools buildings, landscapes and communities.

Costs: Low.

Difficulty: Creating a path for wind to flow is not difficult to understand, but sometimes can be difficult to employ in urban areas because of the many fixed obstructions.

Figure 3.12   **Illustrated above are the types of wind energies that can be captured and utilized in urban areas. These energies are cooling, public health, transportation, wind to work, and wind to electricity.**

Best Scale: Small and large scale. Wind works everywhere.

How-To: The amount of cooling provided by wind is a product of its speed, humidity, and the temperature difference between it and an object or area. Refer to the chapter Thermal Comfort for the specifics of cooling with wind.

Public Health and Air Quality

Uses: Airflow improves air quality.

Costs: Low to high.

Difficulty: Easy to understand but can be difficult to facilitate.

Best Scale: Small and large scale. Wind works everywhere.

How-To: Improving airflow improves air quality in most environments. Refer to the chapter on Public Health for greater detail.

Figure 3.13   **This large breezeway funnels the summer winds and helps cool two courtyards and many classrooms. It has been orientated south-southwest, which is the direction of the afternoon breeze. Moreover, the plants underneath the breezeway are large-leafed and further help to cool the area through evapotranspiration. Picture taken at California Polytechnic University, Pomona.**

Transportation

Uses: Used to move people and goods across ice, land and water.

Costs: Low to high.

Difficulty: Easy to understand, but there are significant barriers to use.

Best Scale: Small and large scale. Transporting via wind energy makes the most sense on water.

How-To: Developed nearly 9,000 years ago to travel between islands, sailing has gone from the primary means of moving goods and people great distances to a recreational pursuit. However, and in terms of energy, sailing is still the most efficient way of travel on two-thirds of the planet.

Wind can both push and lift a vessel. Wind acts on a kite or wing to create motion. Modern sailboats can travel faster than the speed of the wind. The fastest craft sailed on land is Horonku, which broke the record on February 24, 2023, on Lake Gairdner, South Australia. Horonku achieved 140.17mph in winds that were averaging 23mph.

ACTIVE WIND ENERGIES

These are wind energies that require manufacturing, and large amounts of embodied energy. The two most common are turbine to electricity and turbine to process.

Turbines to Electricity

Uses: Generate electricity.

Costs: Medium to high.

Difficulty: Easy to understand but requires technical experience to build and install. Fortuitously, electrical turbines, and mechanical turbines, can be constructed from salvaged urban material.

Best Scale: Small and large scale. Site suitability rules wind energy and urban areas often lack these ideal sites (see below).

How-To: Electricity from wind is easy. Wind spins blades that are attached to a generator and electricity is produced. Turbines typically start producing decent electricity with wind at 7mph; they become economically viable at 9mph; and they begin producing terrific ERoEIs at 13mph and beyond. Turbines are typically braked at wind speeds exceeding 45mph.

Electricity from wind is the best renewable source of energy and large turbines can sometimes repay their economic and energy debt within 6 months.[26] Wind energy, like most active renewable energies, benefits from economies of scale: bigger is better. Wind power output grows exponentially to blade size and tower height.

Site Suitability

Good Wind: Ideally, the wind should move at 13mph for at least 7 hours a day and throughout the year.

Open Areas: Wind becomes more turbulent when passing over obstructions, which increases system stress and shortens a turbine's life. Open spaces are ideal. According to Dan Chiras, a national leader in wind energy, turbines should be sited based on the 30/500 rule.[27] This rule states that turbines should be 30 feet above any obstruction within 500 feet. Unfortunately, this rule implies lot sizes should be 5 plus acres, the tower must be over 60 feet (considered low), and the land manager has control over the height of buildings and trees.

Local Code: Cities and counties have codes and zoning that restrict wind energy. Code will impact the type of device, height, and methods of construction. Zoning can restrict wind generation because of aesthetic impact, lot size, and/or proximity to wildlife.

Agriculture: Cropland is ideal for wind generation because turbines can reduce some types of predations, whether animals, fungus or insects. Turbines also help keep nighttime temperatures above freezing by mixing static layers of air.[28]

Pros

- Great ERoEIs: Whether the measure is biological, economic or energy, wind electricity provides far better returns on investment than many other active energies.
- Turbines are a breeder energy source. Their energy can produce the energy needed for more turbines.
- It is a people's energy because small systems can be built and maintained by the common person and common tools.
- Wind speeds across the globe are increasing, creating greater opportunities for energy generation.[29]

Cons

- Swinging blades kill birds and bats. However, deaths due to turbines are lower than household cats and the use of fossil fuels, which kills animals through the destruction of habitat and climate change.
- They can disrupt wind patterns and alter surface temperatures, disrupting native habitats.
- They are noisy. Small wind turbines make an annoying whine; large turbines maintain a constant "whoosh".
- Small urban turbines struggle to produce enough energy to offset their manufacturing and disposal cost, creating more waste than benefit.

- Modern industrial turbines are constructed of non-recyclable carbon fiber composites, fiberglass, plastics, and resins that are energy drenched and have no other destination than a landfill.

Turbines to Work (Windmills)

Uses: While used mostly for pumping water, windmills are also used to move air, grind, mill, saw, thresh and winch.

Costs: Low to high.

Difficulty: The process is easy to understand, but design and installation requires technical skills.

Best Scale: Small and large scale. Any place with ample wind and a need of a mechanical process. Unfortunately, modern windmills have not been manufactured for large scale industrial uses; electricity has pushed these processes out of use.

How-To: Employing wind energy to run a kinetic process is more efficient than generating electricity to run that same process. Gears do the work instead of generators, converters and regulators. However, urban areas lack both the resources and need for windmills. Wind speeds are generally lower and less reliable in urban areas because of the number of obstructions, such as buildings.

An average modern windmill can pump 2 to 3 gallons of water a minute in 15mph to 20mph winds, which equals about 150 gallons an hour and 1,260 gallons a day (based on wind blowing for 8 hours a day).[30]

Storing Renewable Electricity

Storing renewable electricity is essential to sustaining a city 24/7. However, chemical and conventional batteries are toxic to mine and manufacture, they consume rare materials, and become toxic waste after their relatively short life. Batteries are a destructive solution to a dire problem—how do we store surplus electricity and then release it when needed? Below are the energy storage systems that work best in urban areas, which include gravitational, mechanical, and chemical devices.

Gravitational (Mechanical)

PUMPED-STORAGE HYDROELECTRICITY (PSH)

Electricity is used to elevate a large amount of water. When released the water spins a turbine and produces electricity. Hurtles to this technology are the great amounts of height and water storage needed for cost-effective generation, both of which are in short supply in urbanized areas. PSH may not be advantageous on small scales. The efficiency of these systems ranges between 50% and 85%.[31]

Hydroelectric dams are different than PSH because the potential energy behind the dam is a consequence of nature, and the rain and/or snowmelt that occurred above it. Hydroelectric is not an urban energy because sources of fast-moving water have already been developed and their rights to use are heavily governed.

SOLID GRAVITY ENERGY STORAGE TECHNOLOGY (SGES)

Energy is used to lift a heavy object and when that object drops the energy is retrieved. The falling object spins gears that drive a generator. SGES systems are not difficult to build, take up far less space than pumped-storage hydroelectricity, are non-toxic, are reliable, and are long-lived. Some of the common devices include lifting blocks in the air, lifting blocks in shafts, and lifting pistons with hydraulics. The efficiency of these systems ranges between 70% and 90%.[32]

Mechanical

COMPRESSED AIR ENERGY STORAGE (CAES)

Energy is used to compress air. When released and expanded the air spins a turbine and produces electricity. Air is stored in such things as caverns and tanks. While less efficient than PSH (below) or batteries, these systems are long lived and produce far better energy returns than batteries in the long term. The efficiency of these systems ranges between 40% and 67%.[33]

FLYWHEEL ENERGY STORAGE (FES)

Energy is used to accelerate a flywheel with little drag or friction, and it is saved as a kinetic potential. When needed the spinning flywheel spins a generator and produces electricity. Of the technologies discussed here, this is the most scalable and can be applicable on large and small projects. The efficiency of flywheels systems ranges between 50% and 80%.[34]

Chemical

HYDROGEN (H_2)

Energy is used to split the water molecule and when the hydrogen and oxygen reunite, energy is retrieved. Hydrogen is stored as a gas or liquid. It has advantages and disadvantages. On one hand hydrogen has high energy density, is transportable and can be used

Figure 3.14   **This house is powered solely by hydrogen fuel, whether the hydrogen is used directly, mixed with biogas, or used to generate electricity. The hydrogen fuel was generated from photovoltaic panels. Picture taken at the Energy Resource Center, Downey, CA.**

for transport, is converted to work more efficiently than most fuels, can be converted to other forms of energy, and it is quiet. On the other hand, it requires an incredible amount of energy to split the hydrocarbon or water molecule, it is highly explosive (high transportation costs), and its efficiency for storing energy ranges between 18% and 46%.[35] Hydrogen production is best at large scales.

ERoEI Discussion

ERoEI stands for energy returned on energy invested. It is a broad measure and blunt tool used to measure the efficiency of a wide variety of things, such as buildings, food systems and fuels. A regenerative practice or technology will return more energy than was required to acquire that energy.

ERoEI is measured in units of energy, typically BTUs (British thermal unit) or electricity and kWhs (kilowatt hours). The amount of energy, benefit, and/or work from a system is weighed against the energy needed for mining, manufacturing, transporting and utilizing the energy. ERoEIs are good for broad predictions but struggle to account for any one site because of the large variations in benefit, uses, costs and methods of measurement.

With that said, below is a chart that evaluates urban energies based on a variety of measures. There is often a great range in ERoEIs because of the variables mentioned above.[36]

Terms Defined in Graph Defined

Accessibility: This is a measure of how easy it is to get hold of a particular energy or its feedstock. As an example, food can be grown nearly everywhere and is highly accessible, whereas geothermal electricity can only be generated in specific locations.

ERoEI: Energy returned for energy invested.

Table 3.3 An Overview of Urban Energies

Type	Accessibility	EROEI	Environmental Risks	Quality	Reliable Generation	Storability	Transportability
Biodiesel	Okay	.20 to 9.0:1	Moderate/High	High	Yes	Good	Good
Biomass for Electricity	Locational	.50 to 15:1	Moderate/High	High	Yes	Low	Good
Coal	Good	45 to 80:1	High	High	Yes	Great	Great
Compost	Locational	.50 to 4.0:1	Low	Low	No	Okay	Okay
Ethanol	Okay	.50 to 8.0:1	High	High	Yes	Great	Great
Firewood for Heat	Locational	.20 to .70:1	Moderate	Moderate	Yes	Good	Good
Food	Great	.10 to 40:1	Low/High	Low	Yes	Okay	Good
Geothermal	Locational	7 to 9:1	Low/Moderate	Moderate	Yes	Okay	Okay
Human Direct	Great	.20 to 30:1	Low	Low	Yes	Low	Low
Human Electric	Great	.10:1	Low	Low	Yes	Low	Low
Hydroelectric	Locational	50 to 84:1	Low/High	High	Yes to No	Good	Okay
Nuclear	Good	5.0 to 15:1	High	High	Yes	Great	Good
Oil	High	16 to 20:1	High	High	Yes	High	Great
Solar Hot Water	Great	7 to 20:1	Low	Moderate	No	Low	Low
Solar Passive	Great	5 to 30:1	Low	Low	No	Low	Low
Solar PV	Good	2 to 12:1	Moderate/High	Moderate	No	Low	Okay
Wind Electricity	Locational	10 to 25:1	Moderate	Moderate	No	Low	Okay
Wind Mechanical	Okay	15 to 40:1	Low	Low	No	Low	Low

Environmental Risks: This is a measure of the environmental impacts of growing, harvesting and/or utilizing a particular energy. As an example, the impacts of drying clothes are incredibly small, whereas the impacts of PV panels are high because of the mining, manufacturing and disposal of toxic materials.

Quality: This is a measure of the density or output of a particular energy. Gasoline is of high quality (density), whereas compost is of low quality.

Reliable Generation: This is a measure of the reliability of a particular energy. Is the energy always available, intermittingly available, or seasonally/sporadically available?

Storability: This is a measure of a particular energy's ability to be stored for future use. As an example, ethanol has high storability, whereas human energy does not, we use it or lose it.

Transportability: This is a measure of a particular energy's ability to be transported and used in various locations. As an example, biodiesel has high transportability, whereas solar water heaters do not.

Resources

BOOKS

Chiras, Dan. *Power from the Wind: A Practical Guide to Small-scale Energy Production.* 2nd ed. New Society Publishers, 2017.
Chiras, Dan. *Power from the Sun: A Practical Guide to Solar Electricity.* 2nd ed. New Society Publishers, 2017.
Lyle, John Tillman. *Regenerative Design for Sustainable Development.* John Wiley & Sons, Inc., 1994.
Mazria, Edward. *The Passive Solar Energy Book: A Complete Guide to Passive Solar Home, Greenhouse, and Building Design.* Rodale Press, 1979.

WEBSITES

U.S. Energy Information Administration. Can be viewed at https://www.eia.gov/

Notes

1   Dukes, Jeff. "Bad Mileage: 98 Tons of Plants per Gallon," *UNEWS Archive*, University of Utah. October 27, 2003. https://archive.unews.utah.edu/news_releases/bad-mileage-98-tons-of-plants-per-gallon/
2   "Use of Energy Explained." *U.S. Energy Information Administration.* September 19, 2022. https://www.eia.gov/energyexplained/use-of-energy/electricity-use-in-homes.php.
3   Gorton, Sam. "Compost Power! Is it really possible to extract heat from compost to warm your barn, greenhouse or home?" Cornell University Small Farms Program. October 30, 2020. https://smallfarms.cornell.edu/2012/10/compost-power/

4 Szukovathy, Diane. "Exploring Low-Tech Possibilities for Heating Hoophouses with Compost," *The Cut Flower Quarterly*, vol. 22, no. 3, Fall 2010, pp. 22–23. A grower was able to raise the temperature of her 300 square foot hoop house by 3°F with 8.5 yards of compost and radiant heat.

5 "How Much Firewood Do I Need for Winter?". Timber Works Tree Care. October 30, 2020. https://timberworksva.com/how-much-firewood-do-i-need-winter/

6 "How Much Energy Do My Appliances Use?" Smarter Business. November 7, 2020. https://smarter-business.co.uk/blogs/how-much-energy-do-my-appliances-use-infographic/

7 Avallone, Eugene A., et. al, (eds), *Marks' Standard Handbook for Mechanical Engineers*. 11th ed., McGraw Hill, 2007.
   Kent, Douglas. *A New Era of Gardening: A Book on Gardening for Oxygen and a Healthier Atmosphere*, Garden Shed Productions, 2001.

8 "Bike Fuel: MPB (Miles Per Burrito, Beer, Etc...)," Pure Cycles, Los Angeles, CA. November 1, 2020. https://www.purecycles.com/blogs/bicycle-news/92658567-bike-fuel-mpb-miles-per-burrito-beer-etc?utm_campaign=share&utm_medium=facebook&utm_source=sumome

9 Mortimer, N. D., et. al. "Evaluation of the comparative energy, global warming, and socio-economic costs and benefits of biodiesel," November 3, 2020.
   "Section 1: Biofuels: 1.4 Comparison of biofuels to all fuels used for road transport," UK Department for the Environment, Food and Rural Affairs. December 9, 2021. https://www.gov.uk/government/statistics/area-of-crops-grown-for-bioenergy-in-england-and-the-uk-2008-2020/section-1-biofuels#comparison-of-biofuels-to-all-fuels-used-for-road-transport

10 Hill, Jason, et. al. "Environmental, Economic, and Energetic Costs, and Benefits of Biodiesel and Ethanol Biofuels." Proceedings of the National Academy of Sciences of the United States of America, vol. 103, no. 30, August 2006, pp. 11206-11210. https://www.researchgate.net/publication/6947242_Environmental_Economic_and_Energetic_Costs_and_Benefits_of_Biodiesel_and_Ethanol_Biofuels

11 "Fact Sheet – Biogas: Converting Waste to Energy." EESI: Environmental and Energy Study Institute, October 3, 2017. https://www.eesi.org/papers/view/fact-sheet-biogasconverting-waste-to-energy

12 "Biogas FAQs." Biomass World, November 8, 2020. https://www.biogasworld.com/biogas-faq/

13 Dimpl, Elmar. *Small-scale Electricity Generation from Biomass: Part I: Biomass Gasification*. 2nd ed. Federal Ministry of Economic Cooperation and Development, November 2011, p. 12. https://energypedia.info/images/9/93/Small-scale_Electricity_Generation_From_Biomass_Part-1.pdf

14 "Biomass: Essential for California." California Biomass Energy Alliance, November 5, 2020 at http://www.calbiomass.org/general-statement/

15 "Biomass for Electricity Generation." Whole Building Design Guide, U.S. Department of Energy, September 15, 2016. https://www.wbdg.org/resources/biomass-electricity-generation

16 Watson, Donald and Labs, Kenneth. *Climatic Building Design: Energy–Efficient Building Principles and Practice*, McGraw-Hill Book Company, 1983.

17 "Web Soil Surveys." National Resources Conservation Services, U.S. Department of Agriculture, November 10, 2020. https://www.nrcs.usda.gov/resources/data-and-reports/web-soil-survey

18 Seward, Aaron. "Going Underground: Below the Earth's surface, the temperature remains constant year-round. Geothermal heat pumps harness that energy for efficient heating and cooling." *Journal of American Institute of Architects*, April 6, 2011. https://www.architectmagazine.com/technology/going-underground_o

19 Kaplan, Uri, et. al. "Small scale geothermal power plants with less than 5.0 MW capacity." *Bulletin d'Hydrogiologie*. no. 17, 1999. pp. 433–440. https://pangea.stanford.edu/ERE/pdf/IGAstandard/EGC/1999/Kaplan.pdf

20 *Daylight in Buildings: A Source Book on Daylighting Systems and Components*. International Energy Agency, Solar Heating and Cooling Programme, July 2000. LBNL-47493. https://facades.lbl.gov/sites/all/files/daylight-in-buildings.pdf

21 Mead, Nathaniel M. "Benefits of Sunlight: A Bright Spot for Human Health." *Environmental Health Perspectives*. vol. 116, no. 4, April 2008, pp. 160–167. https://www.ncbi.nlm.nih.gov/pmc/articles/PMC2290997/

22 "Solar Explained: Solar thermal collectors." U.S. Energy Information Administration, December 3, 2019. https://www.eia.gov/energyexplained/solar/solar-thermal-collectors.php

23 Giovannelli, Ambra. "State of the Art on Small-Scale Concentrated Solar Power Plants." *Science Direct: Energy Procedia*. vol. 82, December 2015, pp. 607–614. https://www.sciencedirect.com/science/article/pii/S1876610215026570#!

24 Farrell, John. "Report: Is Bigger Best in Renewable Energy?" Institute for Local Self Reliance, September 30, 2016. https://ilsr.org/report-is-bigger-best/

25 Mulvaney, Dustin. "Solar Energy Isn't Always as Green as You Think: Do cheaper photovoltaics providing solar energy come with a higher environmental price tag?" IEEE Spectrum, November 13, 2014. https://spectrum.ieee.org/green-tech/solar/solar-energy-isnt-always-as-green-as-you-think

26 "Top 5 Commercial Renewable Energy Sources." Duke Energy, April 17, 2021. https://sustainablesolutions.duke-energy.com/resources/top-5-commercial-renewable-energy-sources/

27 Chiras, Dan. *Power from the Wind: A Practical Guide to Small-scale Energy Production*, New Society Publishers, 2017.

28 Biello, David. "How Wind Turbines Affect Your (Very) Local Weather: Wind farms can change surface air temperatures in their vicinity." *Scientific America*, October 4, 2010. https://www.scientificamerican.com/article/how-wind-turbines-affect-temperature/

29 Harvey, Chelsea. "The World's Winds Are Speeding Up: The trend contradicts concerns of a "global stilling," with implications for wind energy." *Scientific America*, November 19, 1019. https://www.scientificamerican.com/article/the-worlds-winds-are-speeding-up/

30 Calderone, Len. "Using Windmills to Deliver Water" *Agritech Tomorrow*, March 13, 2018. https://www.agritechtomorrow.com/article/2018/03/using-windmills-to-deliver-water/10595

31 Decker, Kris De. "Ditch the Batteries: Off-Grid Compressed Air Energy Storage." *Low-Tech Magazine*, March 16, 2018. https://www.lowtechmagazine.com/2018/05/ditch-the-batteries-off-the-grid-compressed-air-energy-storage.html

32 Wenxuan, Tong, et al. "Solid gravity energy storage technology: Classification and comparison." *Energy Reports*, vol. 8, supp. 8, November 2022, pp. 926–934. April 29, 2022 at https://www.sciencedirect.com/science/article/pii/S2352484722022211

33 DiChristopher, Tom. "Hydrogen technology faces efficiency disadvantage in power storage race." *S&P Global Market Intelligence*, June 24, 2021. https://www.spglobal.com/marketintelligence/en/news-insights/latest-news-headlines/hydrogen-technology-faces-efficiency-disadvantage-in-power-storage-race-65162028

34 "Energy and the Environment: Energy Storage." U.S. EPA, April 3, 2018. https://www.epa.gov/energy/electricity-storage

35 DiChristopher, Tom. "Hydrogen technology faces efficiency disadvantage in power storage race." *S&P Global Market Intelligence*, June 24, 2021. https://www.spglobal.com/marketintelligence/en/news-insights/latest-news-headlines/hydrogen-technology-faces-efficiency-disadvantage-in-power-storage-race-65162028

"Hydrogen Explained." U.S. Energy Information Administration, Office of Energy Efficiency and Renewable Energy, no date. https://www.energy.gov/eere/fuelcells/hydrogen-fuel-basics

36 Hall, Charles A. S. *Energy Return on Investment: A Unifying Principle for Biology, Economics, and Sustainability*, Springer, 2018.

Hall Charles A.S., et al. "EROI of different fuels and the implications for society." *Energy Policy*, vol. 64, January 2014, pp. 141–152. https://www.sciencedirect.com/science/article/pii/S0301421513003856

ERoI of Global Energy Resources. SUNY–ESF and NGEI, March 2014. https://mahb.stanford.edu/wp-content/uploads/2014/03/EROI-of-Global-Energy-Resources_SUNYNGEI1.pdf

Capellán-Pérezab, Iñigo, et al. "Dynamic Energy Return on Energy Investment (EROI) and material requirements in scenarios of global transition to renewable energies." *Energy Strategy Reviews*. vol. 26, November 2019. https://www.sciencedirect.com/science/article/pii/S2211467X19300926?via%3Dihub

Chapter 4

Food

The desire to grow food is natural and instinctive. Humans have been doing it for about 12,000 years. The amount of data on food production, whether academic or anecdotal, would fill libraries.

This chapter is not about growing a bountiful harvest but creating systems that regenerate health. That is systems that are as good for the environment as for people. This chapter was written with three goals in mind:

1. Cultivating healthy and nutrient rich foods. This means overcoming the challenges of air, soil, and water contaminants.
2. Producing food that protects the environment and does not undermine biological diversity, soil health, and the condition of waterbodies.
3. Growing food with a low energetic and carbon impact.

Of these three goals, the last is the least understood. Typically, urban foods are loaded with industrial energy (see box below). This creates food that is not sustainable, much less regenerative.

To reach these three goals, this chapter introduces the reader to the three types of food systems (active, passive and spontaneous), the strategies that enhance the environmental benefits of each system, and tactics for growing the healthiest urban food. At the end of this chapter is a section that asks an important question: Is food really the best crop for urban areas?

URBAN FOOD PRODUCTION AND ERoEIs[1]

Energy returned on energy invested is a useful measure when examining environmental remedies and strategies. It is included here to illustrate the dynamics of urban food.

In strictly literal terms, a sustainable food system will produce 1 unit of digestible energy for 1 unit of imported energy: An ERoEI of 1:1. The bulk of the energy needed for growing

DOI: 10.4324/9781003369752-4

food comes from the sun. A regenerative food system will produce food with an ERoEI of 5:1, the minimum ratio needed to sustain vibrant city life.

Unfortunately, urban food is energy drenched. The energy use comes from the many car trips, the highly processed water, the imported fertilizers, and the accessories, such as raised beds and micro-irrigation (heavy on plastics). To put this in perspective, industrial agriculture typically produces food with an ERoEI of 1:6. Urban food ranges from 1:4 to 1:10. Urban food production is rarely sustainable, much less regenerative.

Luckily, there are many things we can do in urban areas to increase the energy efficiency of the food we produce and eat. Keep reading.

Nomenclature

Food production nomenclature is voluminous. Below are many of the words used for systems that produce sustainable and regenerative food.

Agroecology: Agriculture systems that employ the principles of ecology to sustain the health of people, soils, and biological diversity.

Agroforestry: Agriculture that conserves, cultivates, and manages trees.

Annual: A plant that completes its lifecycle in one year or less.

Biennial: A plant that completes its lifecycle in two years. Generally, the first year is spent producing leaves, roots and energy. The second year is spent producing flowers and seeds.

Community Gardens: A large garden comprised of many small plots that are assigned to community members mostly for food production. Community gardens vary in size and can be as small as 20 plots or as big as 200 plots.

Community Supported Agriculture (CSAs): People buy shares of a farm and receive their dividend in food. Shares are purchased monthly or yearly. This is a democratic model of food production and enables dynamic relations between consumer and farmer; the better the farmer does the better the consumer does.

Companion Plants: Companion planting is using an assortment of plants to create better outcomes for a primary group of plants. This primary group might be an orchard, residential shade garden, or community food garden. Companion plants are used to help with fertilization, pest control, and/or pollination.

ERoEI: An equation that measures the efficiency of an energy source. This is a good measure of food because food is a source of energy. The equation means energy returned for energy invested.

Ethnobotany: The study and practice of traditional land management and landscape use. Ethnobotany encompasses the rituals of ceremonies, food, medicine, shelter, and the other aspects of maintaining human life.

Feedstock: Plants grown to feed the animals that produce eggs, hides, honey, meat and/or milk. The term feedstock is also used to describe the plants grown for biofuels.

Foraging: Harvesting and utilizing naturally occurring plants. Food, craft items and medicines are commonly supplied by foraged plants. Foraged plants are part of the natural plant cover and will reproduce and spread without the explicit aid of humans.

Gleaning: Gathering food that would otherwise spoil. As two examples, people glean from public parks that have fruit trees or in agricultural fields after the harvesting machines have finished.

Guerrilla Gardening: Growing plants on land that the gardener has no legal right to work on. Guerrilla gardening can be as simple as scattering seeds and hoping for the best to weeding, planting and irrigating a commandeered landscape.

Integrated Pest Management (IPM): IPM is a method of pest control that strives for pest prevention, pest reduction, and pest resilience. IPM recommends damaging chemicals only after landscape improvements have been made, cultural practices have been changed, and techniques for repelling have been employed.

Livestock: Animals grown on a farm for any number of reasons, such as eggs, fiber, hides, meat, and milk. Petting zoo animals are considered livestock.

Organic: Using living or once living materials to meet the fertilization and pesticide needs of a garden. Organic means no synthetic fertilizers or pesticides.

Passive Food Production: A food system where plants are selected to perform a primary function and provide food, a secondary function. Please refer to the detailed description further on.

Perennial: A non-woody plant that lives over 2 years.

Permaculture: Permanent culture is the literal definition. Permaculture advocates for the ecological integration of human settlement and biological wellbeing to create ecological equilibrium, now and well into the future.

Polyculture: A term in contrast to monoculture. The most famous example of polyculture is the Three Sisters, which is the intertwining of corn, squash and beans. Each plant provides benefits to the nearby plant and receives benefits in return.

Shrub: A woody plant that grows no taller than 30 feet.

Silvopasture: Silva means forest in Latin and the term refers to the practice of incorporating grazing animals to improve the health of a landscape and forest.

Spontaneous Foods: Edible plants that naturalize and spread without the explicit aid of humans. Spontaneous plants are often referred to as weeds, invasives, or people-followers.

Subshrub: A plant that has a woody base but fleshy herbaceous branches.

Tree: A woody plant over 30 feet tall.

Vine: Any plant that can grow up on a vertical surface without the aid of a load bearing trunk. A vine can climb up posts, trees and walls, and it does so by adhering, scrambling and/or tendrils.

Strategies for Urban Food Production

People living in urban areas typically get their food from large-scale industrial farms, not urban areas. However, there are many people that pull all their food from their community, whether that be their landscape, their neighborhood, or a nearby farm. They do it using one of three food systems: Active small-scale, passive, and spontaneous. These three systems are examined below.

Figure 4.1   **An agricultural easement within a city of 309,000 residents made this 12-acre farm possible. The food produced is sold to residents at several roadside stands. Picture taken at Manassero Farms in Irvine, CA.**

Active and Small-Scale Food Systems Defined

Definition: Growing food is the primary goal for these urban landscapes.

Active production includes CSAs and farms under power lines, community gardens, raised beds in schools, and home gardens. They are defined by their small size and possibly many shareholders.

Where Found: From 100 acres in the foothills to 40 square feet in residential backyards, active food production can be found everywhere.

ERoEIs: Most of these food production areas are highly accessorized and mechanized, and their energy footprint is often greater than industrial agriculture, which has an ERoEI of 1:6.

Benefits: Growing food close to the people that will consume it is smart. It keeps money and resources within a community, helps foster beneficial land stewardship, and creates closer bonds between the farmer and consumer.

Figure 4.2 **Micro-irrigation, chicken wire, posts, imported nutrients, and volunteers driving many miles to care for the garden means that the vegetables this small farm produces are energy drenched. While this farm is fantastic for community engagement, it is not beneficial for the atmosphere.**

Problems: Food is grown in urban areas for many reasons, such as recreation and community engagement. However, these reasons do not automatically make the food more environmentally beneficial than industrial agriculture. Small-scale farming lacks the economies of scale of large farms and often produce food with more imbedded energy. Moreover, many of these farms are heavily accessorized, which tends to drive up the energy and carbon costs. Whether backyard or a 50-acre farm, most of these small farms depend on fossil fuels and nutritional imports.

Strategies for Active and Small-Scale Food Systems

Intentionally growing food is a command/control type of relationship. People command the land to produce food and then control the will of nature with an array of tools. This command/control relationship has caused a slew of environmental problems, from excess energy use and carbon emissions to excess fertilization and the degradation of water systems.

This section seeks a closer connection to nature. It examines:

1. Strategies to ensure environmental benefit.
2. Two useful plant lists for urban areas.
3. Cultivating honeybees in urban areas.
4. Animals that can be raised in urban environments.

STRATEGIES FOR ENVIRONMENTAL BENEFIT

There will always be a cost to growing food. The amount depends on the crop, how it is grown, its degree of processing, and amount digested. These costs can be measured in a variety of ways, such as economic, biological diversity, gaseous emissions (carbon, nitrogen, sulfur), water contaminants, energy use, finite resources, labor, topsoil loss, and water use.

The seven recommendations below are aimed at reducing environmental costs, which are primarily air, soil, and water degradation.

1. Design for Consumption: There is no point in growing food if it is not consumed. The whole goal of urban farming is to replace industrial farmland with less harmful practices. Creating relationships and rituals of consumption are as important as growing food.

Designing for consumption means providing the space and opportunity for it. Food producing landscapes should have an outdoor sink and a cutting table for the cleaning, processing, and pre-cooking preparation. Two or three burners would boost consumption. A small shed for drying might also increase rates of consumption. And naturally, seating opportunities makes people feel comfortable about spending time in a space. As a rule, between 2% and 5% of a food producing area should be designated for preparation and consumption.

Figure 4.3   **Harvest and reciprocation, consumption and growth, these are the topics taught at the Lyle Center for Regenerative Studies at California Polytechnic University, Pomona. Pictured is an end of semester harvest meal.**

2. Stack Benefits: Some crops can provide multiple benefits, not just food. When these other benefits are factored into the cost of growing the food, per unit costs fall quickly. For instance, food production can be coupled with urban cooling, personal health and waste handling, creating impressive ERoEIs. Refer to the next section Passive Food.

3. Preserve Food: Since the primary goal of growing food crops is to consume it, preserving food for future use is practical and sustainable. Some of the forms of preservation—fermentation in particular—have the bonus of health benefits. Below are five common ways to preserve food.

- Drying.
- Fermentation.
- Refrigeration (the most energy intense).
- Salting.
- Smoking.

4. Make Own Fertilizer: Importing organic material and chemical fertilizers to grow food is robbing Peter to pay Paul. An urban lot becomes nothing more than a thoroughfare for minerals and nutrients generated elsewhere. There are two ways to generate fertilizer onsite: Grown or compost.

1. Grown: Green crops infuse soils with nitrogen and biomass, which helps create biologically active and nutrient providing soils. If growing on marginal land, then about 20% of the production area is given to nutrient providing crops. These crops rotate throughout the site. Refer to the Fertilizer and Mulch sections in the Landscape Materials chapter for greater detail.

2. Compost: Composting animal byproducts, food waste, and greenwaste will produce nutrient providing and soil building compost and humus. Animal products, such as blood, bone, dairy, egg, manure and meat, should be actively sought, composted and returned to the soil. Bioaccumulation makes animals a powerhouse of macro and micronutrients.

5. *Go Big*: Urban food systems are often more energy consuming than conventional agriculture because they lack economies of scale. The energy used in one truck trip for supplies has a completely different energy impact if a lot is 500 square feet compared to 20,000 square feet—the larger lot can produce 40 times more for the same trip. The lesson is that the larger the production area, the lower the per unit energy costs.

6. *Avoid Using Fossil Fuels*: The goal of a regenerative food system is to have the energy imported to grow a crop match the amount of digestible energy harvested. The goal is 1:1. Achieving this goal is difficult if fossil fuels are used. Fossil fuels are incredibly energy dense and just one gallon of gasoline has the energy equivalent of a person's caloric needs for work for 30 working days. Statistically, the largest energy expenditures are driving people to the landscape and driving to pick up materials.[2]

7. *Honor Spontaneous Foods*: Harvesting spontaneous food is advantageous. They boast fantastic nutrition and have a wide array of health benefits. Many of them are superfoods. They require little or no energy or attention on our part. And serendipitously, they have followed us around the world. Refer to the section on Spontaneous Foods further on.

Figure 4.4   There are many things we can do to reduce the environmental impact of urban food production. Shown in this illustration are four of the many strategies, which are eating the food, employing passive food systems, generating your own fertilizers, and honoring spontaneous foods.

TWO PRACTICAL PLANT LISTS

There are many edible plant lists and hundreds of food plants available in urban areas. Below are two lists of plants that are practical and productive but tend to be under-utilized.

Shade Tolerant Annuals and Herbs

Urban landscapes are commonly shaded for up to 50% of the day by homes and buildings, shrubs and trees. Growing sun-loving plants in shade will reduce quantity and quality of food and increase the amount of disease, fungus and insect infestations. Below are food plants that produce well with half-day shade.

Annuals and Perennials

Arugula, bean, beet, broccoli, Brussels sprout, cauliflower, celery, chives, cress, endive, garlic, kale, leek, lettuce, lovage, parsley, parsnip, pea, potato, pumpkin, radish, rhubarb, rutabaga, salad burnet, salsify, sweet violet, and turnip.

Herbs

Angelica, borage, caraway, catnip, chervil, coriander, lemon balm, lemon verbena, mint, scented geranium, tarragon, sweet violet, and thyme.

Drought Adapted Food Plants

As the planet warms, cities grow, and water managers continue to over-allocate water supplies, water becomes much less abundant. Mandatory rationing is becoming the norm. Below are the food plants that are drought adapted. Also see the section on Spontaneous Foods.

Annuals and Perennials

Amaranth, artichoke, arugula, basil (African), bean, chicory, fennel, mustard, onion, pepper, quinoa, radish, squash, sunflower, tomatillo, tomato, and most fragrant herbs.

Trees

Carob, cherry (native), citrus, date palm, elderberry, fig, guava, honey mesquite, jujube, loquat, mulberry, oak, olive, pomegranate, strawberry tree.

Vines

Blackberry, grape, passion fruit, gourds, kiwi, Malabar spinach, nasturtium, tomato.

Figure 4.5   **Fennel, which is a superfood and fantastic for human circulation and digestion, is easy to grow and propagate. It requires little if any care.**

HONEYBEES IN THE CITY

Cultivating honeybees is a fantastic idea in urban areas. Honeybees produce two valuable crops: Honey and wax.

Honey is more than a super food—it is a sustainer food. Honey has everything humans need. It is packed with energy and nutrients and can improve brain and vascular health. It is also a cough calmer, skin healer, and stress reliever. Bee's wax is a valuable commodity that is used for balms, candles, cosmetics, and soap.

Honeybees are ideal for urban areas. It's not that a gardener will have a great deal of success—rates of colony collapse range from 40% to 70%—but because the nectar and pollen would go to waste if not for trying. And that's the neat thing—everyone should keep trying. Honeybees are non-native and any type of dieback will neither impact biological diversity, food production, nor agricultural endeavors; honeybees rarely fly further than 3 miles from their hives.

However, urban landscapes that border open areas should **never** cultivate honeybees. They compete with native bees for limited resources and will reduce their numbers. Considering that our planet is amid a mass extinction event, gardeners must think and act differently near the wildlands. Honeybees should not be within 5 miles of wildlands.

CHARACTERISTICS OF A VIABLE HIVE

Cultivating honeybees for honey in urban areas is often a good idea. Creating the space for a viable hive requires certain characteristics.

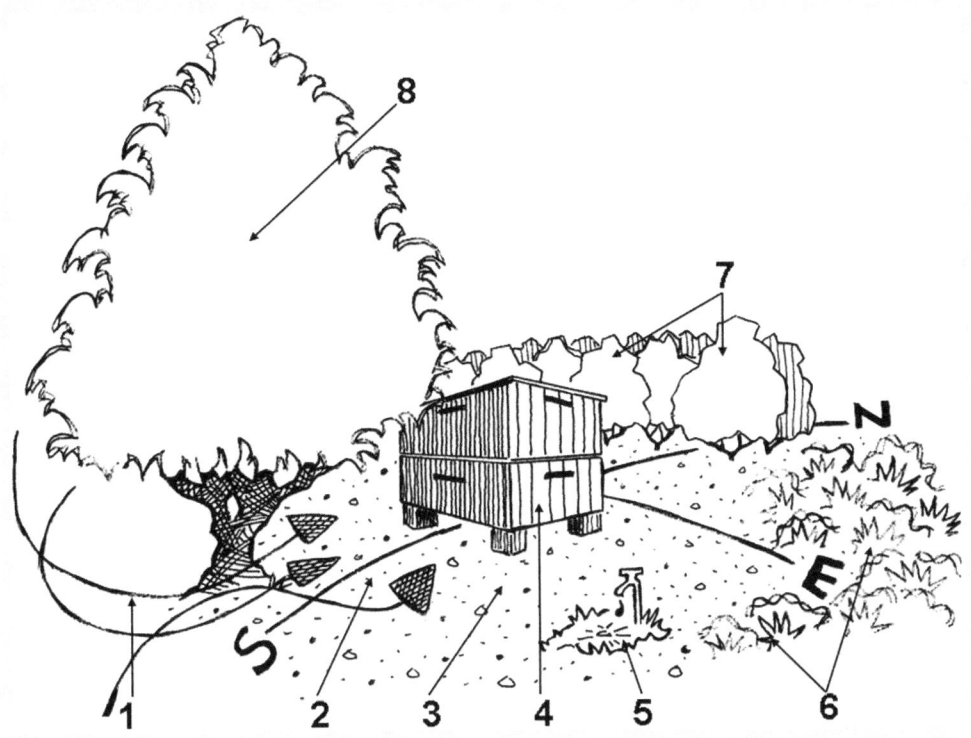

Figure 4.6   1. Air circulation is vital. The cool summer winds should not be blocked. 2. The hive should sit on well-draining and dry soil. 3. Access for harvesting and maintenance must be designed and maintained. The area around hives should be clear of everything but low-lying vegetation. 4. Hives should not be shaded from the morning sun. 5. A water source should be nearby. 6. Foraging areas where nectar and pollen can be found should be within the vicinity. 7. Harsh strong winds should be blocked with a shelter strip. 8. Hives should have filtered sun in the afternoon. However, more sun is always preferrable to more shade. A dry hive is healthier than a damp one.

UTILIZING ANIMALS IN URBAN AREAS

Weaving animals into the urban fabric can produce rich rewards. But to realize those rewards typically means overcoming three hurdles:

1. Too much noise.
2. Too much smell.
3. Too much traffic.

The areas that overcome these problems have the right property size, zoning ordinances, community sentiment, and availability of essential resources, such as feedstock and veterinarians.

If the above hurdles can be overcome, the benefits of adding animals to the urban environment are plentiful. Here are some of their harvestable qualities.

Food: Broth, egg, honey, milk and meat. Some of the best urban food animals are bees, chickens, ducks and rabbits.

Food for Other Animals: Growing animals for other animals to eat is ideal for urban areas. They do not require a lot of space, they do not make much noise, and many can be grown indoors. Some of the animals grown for food include crickets, grubs, mice, snails, and almost any type of larvae.

Fertilizer: Animals are bio-accumulators, and their parts make good fertilizers when processed, either through composting or drying and crushing. Whether blood, bone, feathers, organs or hair, all parts of an animal can be turned into fertilizer and regenerated into a good or service.

Product: Animals produce many valuable products. They can be a source of dyes, such as cochineal; fibers, such as alpaca, goat and sheep; insects for biological pest control, such as lacewings and ladybeetles; and hides, such as goat and rabbit.

Public Engagement: Besides cats and dogs, urban areas have a dearth of animals. Fortunately, most people enjoy animals and creating spaces for the public to interact with them can be profitable. Urban animals are commonly used for petting zoos and pony rides.

Grazing Land Management: If properly managed, grazing animals can provide cost-effective vegetation management. From fire protection to habitat restoration, herbivores are a historic force with many practical outcomes. The most common urban grazers are cows, deer, goats, and sheep. Cows and sheep prefer grasses and forbs. Goats prefer broad-leaf plants less than 4 feet high. Deer eat grasses, broadleaves, and nuts.

Work: Draft animals can be an integral part of a sustainable farming system. Drafting is when you hook an animal to a contraption that drives a process. Some of the uses for draft animals include grinding, hauling, mowing, plowing, seeding, and transportation. The best draft animals are horses, mules and oxen, although goats and sheep are sometimes employed, too.

Passive Foods Defined

Passive food systems should rule. They provide multiple benefits and can easily be woven into every nock and cranny of urban areas. However, this food comes with a caveat: If the community does not harvest and use the food, the system fosters public health hazards in the form of mosquitos, vermin and plant killing pathogens. Creating an ethic of harvest and use are vital to the success of passive food systems.

Definition: A food system where plants are selected to perform a primary function and provide food as secondary function. Primary functions include biological diversity,

Figure 4.7   **Interacting with animals is fantastic for personal health and emotional wellbeing. Unfortunately, few urban residents can smell and touch animals, unless they are cats or dogs. Pictured are the Pony Rides at Griffith Park in Los Angeles, which have created lasting memories for thousands of children for over seven decades.**

cooling/shading, erosion control, privacy, stormwater management, greywater utilization, carbon capture, air pollution removal, and enhancing property and community aesthetics. Whether or not the food is harvested is secondary to the plant's ability to fulfill its primary function.

Where Found: Passive food systems, whether intentional or not, are found throughout the U.S. In fact, the unused food these systems produce has led to new urban movements, such as urban foragers, gleaners and wildcrafters.

ERoEIs: .40 to 10:1. ERoEIs vary due to the amount of processing the food requires and the plant's ability to perform its primary function. A fruit tree, such as avocado, can cut indoor cooling costs by 25% in Southern California. Its ERoEIs are fantastic because not only does it reduce indoor energy use, but by supplying a highly nutritious fruit it eliminates the need for lengthy and environmentally costly transportation from another country.

Benefits: Getting multiple benefits from a plant or system is the key to a regenerative system. Passive food systems are fantastic for urban areas because they provide other essential benefits. Cooling and privacy, biological diversity and improving property values are as important in urban areas as food production.

Figure 4.8   **Pictured is effective intercity cooling in Southern California. Along with dramatically reducing indoor and outdoor temperatures, this avocado tree provides nutritious food, removes gaseous pollutants, produces no notorious allergens, diffuses rainwater, and creates a safe space for children to climb and explore.**

Problems: By far the largest drawback with passive systems is uneaten food. If the food is not eaten or collected, it attracts unwanted animals, diseases and pests, sometimes becoming a public health hazard. Some of the urban animals attracted to fallen food include beetles, mice, mosquitos, opossums, parrots, racoons, rats, and wasps.

Strategies for Passive Food Systems

Below are nine examples of combining service with food production.

PROVIDING COOLING/SHADING

Cooling urban areas is one of the most significant things urban landscapes can do. Plants that cool an area do so in two ways: By providing nearly complete shading and through evapotranspiration. Some of the best food trees for cooling include avocado, camphor, carob, lemon, mulberry, oak (some), olive, pine nut, sapote, tea tree, and walnut. Refer to the chapter on Thermal Comfort for greater detail on cooling.

REDUCING AIR POLLUTION

There are many food plants that can reduce gaseous and particulate pollution. Plants that transpire a lot reduce gaseous pollution. They include avocado, banana, carob, corn, hoja santa, sugarcane, and sunflower. Plants that remove particulates include cedar, cypress and juniper, and they provide berries, fiber and medicine. Refer to the chapters on Public Health and Self-Care for greater detail.

IMPROVING BIOLOGICAL DIVERSITY

There are a great many plants that enhance biological diversity, which in urban areas only means bees, birds, and butterflies. Below is just a handful of the many plants that can attract those species.

- Herbs: Basil, catnip, chamomile, chives, dill, mint, mustard, onion, rosemary and thyme.
- Native: Every state has native plants that provide rich food—for us as well as wildlife. In my state, California, some of the best plants are blackberry, cattail, elderberry, grape, island mallow, Manzanita, mesquite, oak, prickly pear, *Ribes*, rose, strawberry, yarrow, yucca, and wild cherry.

Figure 4.9   **Pictured is a small residential herb garden that is a boon for pollinators. Every plant pictured is animal pollinated, easy to grow, and most can be used in a kitchen. Basil, lavender, mint, rosemary, sage and thyme flourish.**

SEQUESTERING CARBON

Ninety percent of the dry weight of a plant, called biomass, is comprised of carbon dioxide pulled from the atmosphere. A carbon sequestering plant, therefore, is one that accumulates a lot of biomass. Carbon capturing plants are both fast growing and long lived. Some of the best plants for food and sequestration include almond, avocado, Brazil nut, carob, cashew, hazelnut, macadamia, oak, pecan, pine nut, pistachio, sapote, sugarcane, and walnut. Pasture plants for livestock, which is a mixture of annual and perennial grasses and forbs, are good for sequestration too, but the carbon is stored in the soil.

FIRE PROTECTION

Annual, perennial and trees crops have been used for centuries to protect structures and communities from wildfires. Landscaping for fire protection hinges on a high degree of maintenance. Food crops ensure this diligent degree of maintenance. Almost any type of food crop will work, assuming they are kept free of fine ignitable material and have proper leaf moisture.

CONTROLLING TOPSOIL LOSS

Topsoil loss in urban areas undermines landscape health of stormwater infrastructure, and waterbodies. The plants that reduce topsoil loss lie on top of soil and protect it from water and wind erosion. Some of the hill hugging food producing plants include *Berberis*, blackberry, dayflower, Manzanita, natal plum (prostrate), oregano, nasturtium, New Zealand spinach, rose, rosemary, sagebrush, saltbush, sea fig, sea rocket, and strawberry.

Figure 4.10   **Food has saved communities and homes from wildfire for hundreds of years. The citrus pictured here protected a Southern California housing tract from a conflagration that devoured hundreds of acres. Picture taken in Yorba Linda, CA.**

UTILIZING GREYWATER

Irrigating with greywater is different than other waters. Not only does it have more salts, nutrients, and alkalinity, but it is also delivered daily. The food plants that thrive with greywater are different too; they can tolerate salts and possibly saturated conditions. Along with the list below, refer to the chapter on Greywater.

- Annuals and Perennials: Artichoke, asparagus, cucumbers, peppers, tomatoes, zucchini.
- Berries: Blackberry, blueberry, hop, raspberry, strawberry.
- Deciduous Fruit: Almond, fig, grape, plum, persimmon, walnut.
- Herbs: Comfrey, lemon balm, lavender, mint, rosemary, sage, yerba mansa.
- Tropical: Banana, cherimoya, date palm, guava, mango, passion fruit.

PRIVACY

Creating privacy is an essential need in urban areas. Plants that create this comfort are dense but not too tall, easy to grow but not too fast growing. Some of the best plants include banana, bay, cherry, citrus, elderberry, guava, Chinese jujube, laurel, olive, pineapple guava, *Ribes*, serviceberry, strawberry tree, sumac, sugarcane, wax myrtle, and vines such as blackberry, grape, hop, kiwi, passion fruit, and raspberry.

WASTE HANDLING

Everybody produces waste and returning these nutrients back into the food system is foundational to a sustainable food system. Sludge from reclamation plants, detritus from livestock, and coarse mulch are all ideal soil amendments. The plants capable of making the most of this nutrient rich matter include fast growing annuals, biennials, temperate shrubs and trees, and many tropical plants.

Spontaneous Foods Defined[3]

Americans need to change their attitude towards spontaneous foods, or as most of us think of them weeds. Not only are the majority of urban weeds edible, but many are also considered superfoods. Instead of waging our decades old war against these phenomenally nutritious plants and replacing them with plants with mediocre benefits, we should accept them, embrace them, and most of all consume them.

In fact, weeds are so beneficial that many can be found in grocery stores or as decoctions in health stores. These plants include blackberry, black walnut, blue gum *Eucalyptus*, carob, camphor, chicory, dandelion, daylily (flower), elderberry, fennel, fig, grape (leaf and fruit), lemon-scented gum, mint, nasturtium, nopales (prickly pear), passionfruit, purslane, snail, stinging nettle, strawberry, and watercress.

Below is an overview of spontaneous foods, strategies to protect yourself and the source of food, and a list of 22 amazing commonly found foods.

Definition: Spontaneous food systems are naturally occurring or human caused, but with minimal, if not unintentional effort. The practice of harvesting spontaneous food encompasses these pursuits: Ethnobotany, foraging, gleaning, guerrilla gardening, and wildcrafting.

Where Found: Spontaneous foods can be found everywhere, even parking lots. The practice involves harvesting the naturally occurring goodness.

ERoEIs: 3.0 to 10:1. The embodied energy of spontaneous systems comes from the bounty of life – our mother's sun, soil, water, and air. Our invested energy is tiny. We find it, we pluck it, and we eat it. ERoEIs are fantastic. Notably, if there are big game animals, and the community could eat them, the ERoEIs could climb to 60:1; the energy supporting these animals coming from vast tracts of open space and its ability to accumulate digestible biomass and energy.

Benefits: Of all the urban food systems, spontaneous foods not only have the lowest energy footprint, but might also be the healthiest. Keep reading.

Figure 4.11 **Carob is a tree that follows humans. They love us, and in ancient times, gardeners loved them. They produce a seedpod that is a superfood. It is loaded with protein, sugar, and essential nutrients. The trees also provide dense shade and are evergreen, protecting us in summer and winter.**

Problems: There are risks with harvesting from unfamiliar areas, which are mostly digestive and involve pathogens and toxins. Luckily, the protocols to overcome the risks are easy to understand and follow.

Strategies for Spontaneous Food

Below is a brief overview of working with spontaneous foods. It includes how to protect yourself and the harvest plus a list of some of the most valuable foods to forage.

PROTECTING YOURSELF

While harvesting spontaneous foods is not dangerous, it is not without risk. There are many threats, whether digestive, physical, or legal. Below is a quick overview of overcoming digestive risks.

Eat Only What You Can Identify: Never, ever, eat some part of a plant unless the plant has been positively identified as edible. There are poisonous plants that look scrumptious.

Slow Down: Never rush in and harvest wild food. Slow down and pay attention to the environment. Impatience leads to bug bites, thorn pricks, food with fungus, trampling young food, and the tragedy of overharvest.

Identifying Herbicides: Do not eat wild foods if pesticides are suspected. While declining, pesticide use is still prevalent in urban areas. Identifying contrasts is the secret to determine whether or not herbicides have been used in an area. Herbicides always leave a tell-tale line between the treated areas and not.

Reducing Metals: Industrial processes and transportation release fine metals into the environment. Digesting metals is terrible for personal health. Avoid wild foods within 500 feet of an industrial area and highways. Importantly, always wash wild foods in water warmer than the temperature of the item to ensure that toxins are pushed out and not pulled in.

Overcoming Pathogens: If concerned about the pathogens, boil the food. Taking the food to 165 degrees ensures cleanliness. Flashing boiling is a good strategy for plants that rapidly lose their nutrients to heat, such as baby greens.

PROTECTING THE HARVEST

There are people that do not treat the land and harvest with respect. For instance, people might overharvest or trample an area, leading to no food for other people and wildlife. Follow the guidelines below to ensure a harvest now and well into the future.

Never the First: Harvesting the first found wild food is as bad for you as it is for the species. Grabbing the first found could mean that the species dies out, or that the harvest is not the healthiest, or that you undermine another plant and its health and harvest. Always survey a broad area before choosing.

Never More than 50%: Many spontaneous plants grow because native plants struggle to compete. The environment has changed too much. They take advantage of an opportunity and provide some type of service, whether slowing erosion, providing pollen and nectar, or removing pollutants from the air or soil. Restricted harvesting allows these other needed services to continue.

Let It Spread: Whether it is by seed, rhizome, or stolon, letting a plant roam is vital. Naturalization is nurtured to ensure a harvest now and well into the future.

Improve Conditions: Reciprocation goes a long way, even if it is just a little. Pull smothering plants, snip shading plants, brush off pests, pee around plants, and remove diseased parts to ensure health and a harvest the following year. Give and take, that's the deal.

TWENTY-TWO NUTRITIOUS FOODS THAT CAN BE FOUND IN URBAN AREAS

Below are the plants, and one animal, that can be found in urban areas throughout the U.S. They also can be cultivated at little cost, as measured in energy, money, resource use, time, and water. In most cases the species below are better for personal health than the foods found in grocery stores.

Blackberry, *Rubus* spp.: Berries and leaves are edible. The berry is rich in fiber, manganese and vitamins C and K, and is an antioxidant and anti-inflammatory. It also has carbohydrates and energy.

Black mustard, *Brassica nigra*: Leaves, flowers and seeds are edible. Black mustard is a superfood. It is very high in fiber and vitamins A, C and K. It also has a lot of calcium, folate, iron, manganese, potassium, and vitamins B6 and E. And it has some fatty acids, protein, and phosphorus. Like many mustards, black mustard is anti-inflammatory and antioxidant.

Cat's Ear, *Hypochaeris* spp.: Leaves, stems and flower buds are edible. Cat's ear is rich in antioxidants, fiber, and many nutrients.

Cheeseweed, *Malva parviflora*: Leaves, stalks, fruits and roots are edible. Cheeseweed is a superfood. It is rich in chlorophyll, calcium, fiber, iron, magnesium, pectin, potassium, selenium, and vitamins A and C. Its leaves, shoots and roots are anti-inflammatory and antioxidant.

Chickweed, *Stellaria media*: Leaves, stems and seeds are edible. Chickweed is a superfood. It is high in beta-carotene, calcium, chlorophyll, iron, magnesium, potassium, riboflavin, vitamin C, and zinc. Chickweed is an anti-inflammatory, antioxidant and promotes lymphatic response.

Chicory, *Cichorium intybus*: Leaves and roots are edible. Roots are packed with starchy energy and have some sugar and protein. Leaves provide iron and vitamins B, C and K. Chicory is an antioxidant.

Curly dock, *Rumex crispus*: Eat the abundant leaves. Curly dock is high in vitamins A and C, iron, and potassium. It also has some protein.

Dandelion, *Taraxacum* spp.: Entire plant is edible: Leaves, stems, flowers and roots. Dandelion leaves are a superfood. They are loaded with calcium, essential fatty acids, iron, magnesium, phosphorus, potassium, sodium, and vitamins A, B and C. The entire plant is both anti-inflammatory and antioxidant.

Fennel, sweet, *Foeniculum vulgare*: Young leaves, stems, bulb, and seeds are edible. Fennel is a superfood. It is an excellent source of vitamin C. It is a very good source of copper, fiber, folate, manganese, phosphorus, and potassium. And it is a source of calcium, iron, magnesium, and vitamin B3. Fennel is anti-inflammatory and an antioxidant.

Fig, common, *Ficus carica*: The fruit of this small tree is eaten raw, dried, brewed or cooked into jams and sauces. Figs are a good source of calcium, copper, iron, potassium, sugar and vitamin A.

Figure 4.12 **Humans are evolved to digest every part of a dandelion. The leaves, stalks, flowers and roots are all edible and packed with vital nutrients and medicines. Dandelions are more than a superfood.**

Filaree, red-stemmed, *Erodium cicutarium*: While it is mostly harvested for leaves, the entire plant is edible. Its leaves taste like parsley. Filaree contains many vital nutrients, vitamins and fiber, and has antioxidant qualities.

Goosefoot and lambsquarter, *Chenopodium album* and *C. murale*: Eat the leaves and seeds. These plants are nearly a superfood and contain high amounts of calcium, manganese and vitamins A and C, along with some B6, protein and riboflavin.

London rocket, *Sisymbrium irio*: Leaves, flowers and seeds are edible. London rocket is rich in nutrients, vitamins, and roughage. It is an antioxidant and anti-inflammatory.

Miner's lettuce, *Claytonia perfoliata* (and subspecies): Leaves and stems are edible. Miner's lettuce has kept adventurers and pioneers healthy for hundreds of years. The leaves are high in vitamin C and have beneficial amounts of iron and vitamin A.

Nasturtium, *Tropaeolum majus*: Leaves, stems, and flowers. It contains fiber and many nutrients and vitamins. Nasturtium also has antioxidant qualities. Its flowers are high in vitamin C.

Plantain, *Plantago major*: Eat the abundant leaves. Plantain is nearly a superfood. The leaves are very high in calcium, fiber, and vitamin A. They are also high in vitamins C and K. They have mild anti-inflammatory and antimicrobial properties and are excellent as a salve for healing wounds.

Purslane, *Portulaca oleracea*: The leaves and stems are edible. While not usually listed as a superfood, many people claim it is. It is high in all the essential nutrients, including iron, omega-3s and vitamin E.

Rose, *Rosa californica*: The hips and flower petals are edible. Hips have been used for centuries for colds, pain, and many other maladies. It is high in minerals and vitamins, including A and C, calcium, iron and phosphorus.

Sourgrass, *Oxalis pes-caprae*: Leaves, stems, flowers and corns (root nodules) are edible. Sourgrass is high in vitamin C and contains vitamin A. It is high in oxalic acid, which should be enjoyed in moderation. Corns are delicious roasted.

Stinging nettle, *Urtica* spp.: Eat the leaves. Stinging nettle is a superfood. It is rich in chlorophyll, calcium, iron, magnesium, potassium, and vitamins A, C, D and some K. It is an anti-inflammatory.

Wild radish, *Raphanus sativus*: Leaves, stems, flowers, seeds, crown and roots are edible. The leaves and roots are high in fiber and vitamin C and improve digestive function. Radish seeds are a stimulant, and they are consumed either raw or brewed.

Figure 4.13   **Purslane is delicious and lemony, nutritious and healthy, abundant and common. Pictured is purslane added to a bowl of avocado, onion, olive oil, sage, and salt.**

SNAILS

If you trust the soil, there is no reason not to trust eating the snails it encourages, a delicacy that has been cultivated and eaten for centuries. Snails are rich with Omega 3 fatty acids and fantastic for humans. They are also high in calcium, iron, magnesium, phosphorus, and zinc.

There is a common misconception that to eat a snail its digestive tract must be first cleaned with a diet of cornmeal. This is not true. The risk of eating snails are metals and pathogens. Metals are avoided by creating toxin-free soils (see strategies below). Pathogens are cooked out. Below is a simple recipe for preparing snails from healthy gardens and soils.

Live snails should be thoroughly boiled before consumption. In a pot of seasoned water boil the snails for about 8 minutes, change the water and boil again. Snails will produce light scum. They are cooked when they no longer produce scum. After they have been boiled, sauté them with butter, garlic and seasoning.

Growing and Eating Healthy Food

Growing food free of pathogens and toxins is fundamental to health and wellbeing. Unfortunately, urban areas are sometimes rife with pathogens and pollutants. Use the guide below to grow healthy food, no matter how big your plot is.

Healthy Soils

Healthy soils grow healthy people. Everybody knows that. What urbanites don't know, however, is what condition their own soil is in. What impact did the decades of other uses have on the toxicity of the land? This uncertainty makes people reluctant to grow food.

There is ample reason to be weary of urban soils. Arsenic, cobalt, cooper, chromium, lead, nickel, zinc, and many types of hydrocarbons are a legacy of our industrial endeavors and haunt some urban soils. The toxins and metals found in urban soils mostly come from atmospheric deposition and urban runoff, which is tainted by the impacts of transportation, stack emissions, and construction and industrial wastes. Pressure treated wood, along with exterior and interior paints are also prevalent sources of contaminates.[4]

Yet given all of that, growing food in urban soils may be better for you than soils in rural or agricultural areas, and there are 3 good reasons:[5]

1. **Zoned Out**: Many of the practices that produced metal and toxin rich soils were outlawed and/or excluded from urban areas through better zoning decades ago, the most notable being the ban on lead paint in 1978.
2. **Lowlands**: Many of our cities sit in the lowlands, areas that would have been teaming with water and life for thousands of years, developing deep and rich soils. Human activities also provide macronutrients and urban soils generally have higher amounts of potassium and phosphorus, and in some cases nitrogen, than native or natural soils.[6] There are many great soils in the U.S.
3. **You**: And lastly, you can alter and improve your soil—you cannot alter the soil of industrial agriculture. People, and their powerful energy, can easily create healthy conditions. See the list below for tips for reducing toxins.

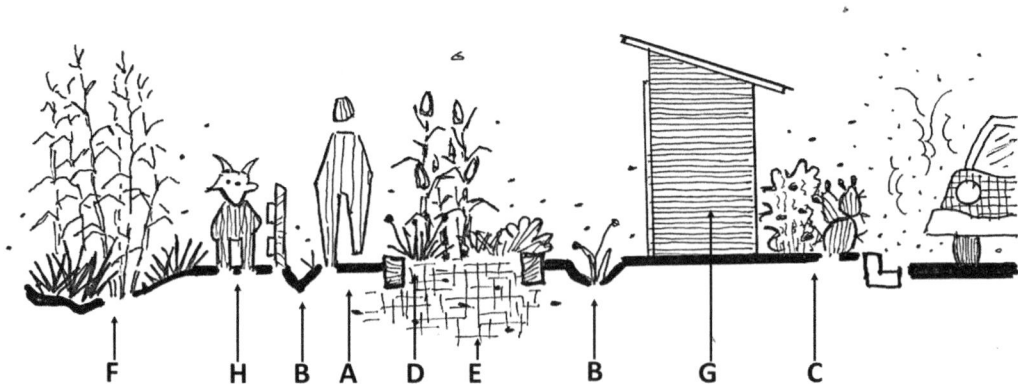

Figure 4.14   **The letters above correspond to the strategies below.**

A. Test soil. B.Divert runoff and sheet flow. C.Eat the extremities of plants. D.Use compost. E.Import and/or enhance soil. F.Phytoremediation. G.Store chemicals away from food production. H.Separate animals from plant production.

STEPS YOU CAN TAKE TO ENSURE THE HEALTH OF YOUR SOIL

Test Soil: Have your soil tested. All you have to do is mail in a sample. No driving involved. Download and fill out a Soil Testing Form and mail your selected samples. Importantly, select the test for Agricultural Soil Suitability Analysis, which identifies the concentrations of 15 non-essential trace metals, which includes toxic metals.

Divert Runoff: The water from roofs, driveways, and streets is known to have harmful metals, pathogens, and other ultrafine contaminants. This water must be diverted from food production areas. Refer to the chapter on Rainwater Capture for greater detail.

Eat From Extremities: If afraid of the toxins in the soil, cultivate foods that grow at the end of a stem, such as flowers, fruits, nuts and seeds. Toxins and metals get held up in a plant's vascular system and the eternities are much cleaner. Leaf and root crops should always be avoided along busy streets and in commercial/industrial areas.

Add Compost: Adding organic matter to unhealthy soils can help in one of two ways. First, it can dilute the toxins, making them less likely to be pulled up by a food crop. Second, organic matter, along with the microbes that devour it, can bind some types of metals and make them unavailable for uptake.

Import and Raise Soil: Soils immediately around buildings have a higher likelihood of metals and toxins. It is the weatherization of the building and human discharge, which impacts the air, soil, and water. If the building is older than 1979, the soil may contain lead because those paints were dominate before then.[7] Importing and raising soil will reduce the impacts of weatherization and improve the cleanliness of food producing soils. Raised beds are best for annuals, biennials, and shallow rooted perennials. Large plants with aggressive and large roots are ill-advised around buildings.

Phytoextraction[8]: There are many plants that can synthesize the metals into their tissue. They are often called hyper-accumulators. Some of the best at accumulating metals come from these genera: *Acacia, Amarathus, Ailanthus, Arundo, Brassica, Cenchrus, Cyperus, Hypercicum, Pteris, Helianthus annuus, Miscanthus* × *giganteus, Nicotiana, Populus, Tamarix,* and *Typhia*.

Unfortunately, phytoextraction is not a quick process. Pulling metals via photosynthesis and biomass creation is slow. Meaningful extraction will take years. Furthermore, the greenwaste from these endeavors must be sent to a landfill or hazardous waste facility. This material should not be composted, as that would return the metals back to the environment.

A GOOD RESOURCE

For more detailed information on good agricultural practices, refer to the state of California's guidelines on growing healthy food at small scale at Good Agricultural Practices (GAP) & Good Handling Practices (GHP) | Agricultural Marketing Service (usda.gov)

Growing Healthy Food Along Streets

If growing crops along a busy or semi-busy street or in an area where the public has access 24/7, avoid leaf and root crops. Cars produce all types of airborne contaminants that can accumulate in leaves and roots. Moreover, gardeners cannot control unfortunate accidents, such as a spill of radiator fluid. The best practice is to grow crops that sit at the end of a stem, like flowers, fruits, nuts and seeds. Toxins and metals will get held up in a plant's vascular system and do not make it to the plant's extremities.

Are Food Crops the Best Crop for Urban Areas?

Urban areas consume resources imported from across the world. Building materials, ceremonial items, medicines, textiles, and timber are in great demand. And because of this, food is not always the wisest crop for urban areas.

While obviously food is very important, realistically it is a commodity that can be grown for less energy and money elsewhere. Large farms have the advantage of economies

Figure 4.15   Artichoke, stone fruits, passionfruit, and citrus grace the front of this small urban farm. A busy street runs in front of it and the organization decided to protect the community from toxins with food that grows at the end of a branch. Picture taken at the Orange Home Grown Education Farm, Orange, CA.

of scale. Food also has many inconveniences: It spoils quickly (eat it or lose it), demands a great amount of time and attention, and can create unwanted vector problems.

Luckily, urban areas have one huge advantage over industrial processes: Urbanites have more money and time to invest per square foot. What this means is that we are great at producing boutique, specialty, value-added, and unusual crops.

Some of the unique crops that might be better than food include building materials, cut flowers, dyes, fibers, medicines, pest control, propagules, and seasonal and ceremonial items. All these other things are discussed elsewhere in this book. Below is a list for herb and tea plants, which is one of the good alternatives.

Plants for Teas

While herbs and teas do not provide many calories, and are not a sustaining food, they do enhance a person's ability to metabolize, utilize, and pass the calories eaten. They are important to good health. Using homegrown ingredients, the freshest, improves their efficacy. Most of these plants are easy to grow. Some will even emerge spontaneously as weeds.

Bay (native and sweet), *Umbellularia californica* and *Laurus nobilis*: Use leaf.

Bergamot, wild, *Monarda fistulosa*: Use leaf and flower.

Black mustard, *Brassica nigra*: Use seed.

Blue gum, *Eucalyptus globulus*: Use leaf.

Camellia, *Camellia sinensis var. sinesis* and *C. sinensis var. assamica*. Use leaf to make black, green and white teas.

Chamomile, *Matricaria recutita*: Use flower.

Chicory, *Cichorium intybus*: Use leaf and root.

Cilantro, *Coriandrum sativum*: Use seed.

Coneflower, *Echinacea* spp.: Use root.

Dandelion, *Taraxacum* spp.: Use flower and root.

Elderberry, *Sambucus* spp.: Use flower.

Fennel, *Foeniculum vulgare*: Use seed.

Ginger, *Zingiber officinale*: Use root.

Ginkgo, *Ginkgo biloba*: Use leaf.

Jasmine, *Jasminum* spp.: Use flower.

Lavender, *Lavandula* spp.: Use flower and leaf.

Lemon, citrus: Use fruit and petal.

Lemon balm, *Melissa officinalis*: Use leaf.

Lemonade berry, *Rhus integrifolia*: Use fruit and leaf.

Lemon grass, *Cymbopogon*: Use leaf.

Lemon gum, *Corymbia citriodora*: Use leaf.

Lemon verbena, *Aloysia citrodora*: Use leaf.

Licorice, *Glycyrrhiza glabra*: Use root.

Mint, *Mentha* spp.: Use leaf.

Marjoram, sweet, *Origanum majorana*: Use leaf.

Mugwort, *Artemisia vulgaris*: Use leaf.

Oregano, *Origanum vulgare*: Use leaf.

Pineapple weed, *Matricaria discoid*. Use flower.

Figure 4.16   Chicory is a common weed and easy plant to grow in urban areas. Its roasted root is a coffee substitute, and its leaves provide high nutrition.

Rose, *Rose* spp.: Use fruit (hip) and petal.
Rosemary, *Salvia rosmarinus*: Use leaf.
Sage, *Salvia officinalis*: Use flower and leaf. Many varieties of sage are used.
Stevia, *Stevia rebaudiana*: Use leaf.
Stinging nettle, *Urtica* spp.: Use leaf.
Strawberry, *Fragaria* spp.: Use fruit and leaf.
Sugarbush, *Rhus ovate*: Use fruit and leaf.
Tarragon, *Artemisia dracunculus*: Use leaf.
Turmeric (wild), *Curcuma aromatica*: Use root.
Thyme, *Thymus* spp. Use leaf.
Valerian, *Valeriana officinalis*: Use root.
Violet, *Viola odorata*: Use flower.

Resources

BOOKS

Clarke, Charlotte Bringle. *Edible and Useful Plants of California*. University of California Press, 1977.
Elpel, Thomas J. *Botany in a Day: The Patterns Method of Plant Identification*. 6th ed., HOPS Press, LLC.,
    2013

Jeavons, John. *How to Grow More Vegetables: (and Fruits, Nuts, Berries, Grains, and Other Crops) Than You Ever Thought Possible on Less Land Than You Can Imagine.* 8th ed., Ten Speed Press, 2012.

Kent, Douglas. *Foraging Southern California: 118 Nutritious, Tasty, and Abundant Foods.* Wilderness Press, 2020.

Markham, Brett. *Mini Farming: Self-Sufficiency on 1/4 Acre.* Skyhorse, 2010.

Nyerges, Christopher. *Foraging California: Finding, Identifying, and Preparing Edible Wild Foods in California.* Falcon Guides, 2014.

WEBSITES

Desert Harvesters. Can be viewed at https://www.desertharvesters.org/

Eat the Weeds: And other things too, Green Deane. Can be viewed at https://www.eattheweeds.com/

Plants for a Future. Can be viewed at https://pfaf.org/

Notes

1 The ERoEIs listed in this chapter came from the following sources:

Qualman, Darrin. "Earning Negative Returns: Energy Use in Modern Food Systems." Graphic Descriptions, August 1, 2017. https://www.darrinqualman.com/energy-use-in-modern-food-systems/

Martinez-Alier, Joan. "The EROI of Agriculture and its Use by the Via Campesina." *The Journal of Peasant Studies,* vol. 38, January 2011. https://www.tandfonline.com/doi/abs/10.1080/03066150.2010.538582?scroll=top&needAccess=true&journalCode=fjps20

Tillman, John Lyle. *Regenerative Design for Sustainable Development.* John Wiley & Sons, Inc., 1994.

Markussen, Mads V. and Østergård, Hanne. "Energy Analysis of the Danish Food Production System: Food-EROI and Fossil Fuel Dependency." *Energies,* vol. 6, August 15, 2013, pp. 4170–4186. https://pdfs.semanticscholar.org/24a7/4c02b99d6c9ecd9fc712e7baa24a08522790.pdf

Lott, Melissa C. "10 Calories in, 1 Calorie Out – The Energy We Spend on Food." *Scientific American,* August 11, 2011. https://blogs.scientificamerican.com/plugged-in/10-calories-in-1-calorie-out-the-energy-we-spend-on-food/

2 Kent, Douglas. *A New Era of Gardening: A Book on Gardening for Oxygen and a Healthier Atmosphere,* Garden Shed Productions. 2001.

3 The phrase Spontaneous Foods was adapted from an article by: Riano, Izabela. "Just A Bunch of Weeds: An Interview With Peter Del Tredici." *Scenario 02: Performance,* Spring 2012. https://scenariojournal.com/article/peter-del-tredici/

4 Pouyat, Richard V., et al. "Chemical, Physical, and Biological Characteristics of Urban Soils." *Urban Ecosystem Ecology,* USDA Forest Service, Northern Research Station, ch. 7, 2010, pp. 119–152. https://www.fs.usda.gov/research/treesearch/36426

and

"Soil Quality Urban Technical Note No. 4: Urban Soil in Agriculture." USDA, Natural Resources Conservation Service. April 2017. https://directives.sc.egov.usda.gov/OpenNonWebContent.aspx?content=41327.wba

5 Zigas, Eli. "Is City Soil Really More Toxic Than Rural Soil?" *SPUR,* September 14, 2011. https://www.spur.org/news/2011-09-14/city-soil-really-more-toxic-rural-soil

6 Pouyat, Richard V., et al. "Chemical, Physical, and Biological Characteristics of Urban Soils." *Urban Ecosystem Ecology,* USDA Forest Service, Northern Research Station, ch. 7, 2010, pp. 119–152. https://www.fs.usda.gov/research/treesearch/36426

Valuing Chaparral: Ecological, Socio-Economic, and Management Perspectives. Ed. Underwood, Emma C.; Safford, Hugh D.; Molinari, Nicole A.; Keeley, Jon E. Springer Nature 2018, p. 159.

Zigas, Eli. "Is City Soil Really More Toxic Than Rural Soil?" *SPUR*, September 14, 2011. https://www.spur.org/news/2011-09-14/city-soil-really-more-toxic-rural-soil

7 "Guidance for San Francisco Residents and Public Agencies: Lead Hazard Risk Assessment and Management of Urban Gardens and Farms." City and County of San Francisco, Department of Public Health and Environmental Health, February 2017. https://www.sfdph.org/dph/files/EHSdocs/ehsCEHPdocs/Lead/LeadHazardUrbanGardening.pdf

8 Tangahu, Bieby Voijant, et al. "A Review on Heavy Metals (As, Pb, and Hg) Uptake by Plants through Phytoremediation." *International Journal of Chemical Engineering*, vol. 2011, January 2011. https://doi.org/10.1155/2011/939161

Ranieri, Ezio, et. al. "Ailanthus Altissima and Phragmites Australis for Chromium Removal from a Contaminated Soil." *Environmental Science and Pollution Research*, vol. 16, August 23, 2016. https://pubmed.ncbi.nlm.nih.gov/27146531/

Pivetz, Bruce E. "Phytoremediation of Contaminated Soil and Ground Water at Hazardous Waste Sites." National Risk Management Research Laboratory Subsurface Protection and Remediation Division, Environmental Protection Agency, February 2001. https://www.epa.gov/sites/default/files/2015-06/documents/epa_540_s01_500.pdf

Chapter 5

Landscape Materials

Capturing and growing landscape materials onsite will reduce greenwaste, allow a landscape to store more carbon, and offset the upstream costs of importing materials. On a personal and physical level, generating landscape features, such as fences and walls, onsite is both physical and rewarding.

Below are many of the landscape materials that can be generated onsite with nothing more than a chainsaw, chipper, hand pruner, machete, and/or shovel. Covered are earthen products, fertilizers, fibers, mulch, pest control, propagules, screens, and woody products, such as planks and poles.

Quick Overview of Making Landscape Materials

Uses: Products that can be generated include earthen products (foundations, siding, steps, walls), edging, fibers (cordage, weaving), handrails, logs (benches, steps, walls), mulch (mulch, compost, humus), pesticides, planks, poles, propagules (new plants), stakes, screens, walking surfaces, and wattle (erosion control, screens, weaving).

Costs: Low to moderate.

Difficulty: Budgeting the effort and time are difficult, not the processes.

Best Scale: Big landscapes with big vegetation and big needs are the best candidates for generating/using landscape materials. This would include nature centers, environmental educational facilities, city and county parks, state and federal recreation areas, and large residential landscapes. While all landscapes can pursue some part of this chapter, not all can absorb all the material that their landscape produces.

DOI: 10.4324/9781003369752-5

Figure 5.1   **Everything in this picture, which includes the bench, posts, beams, roof, privacy screen and basket, were created from vegetation within 500 feet of it. Picture taken at the Marin Art and Garden Center, Ross, CA.**

ERoEI: Generating a site's landscape materials can provide good energy returns because it avoids transporting waste off and new materials on. Also, many of these materials can be processed and prepared with hand tools, avoiding industrial energy and machinery. However, the economic returns may never be as beneficial as the energetic and environmental returns because of the leap in labor costs.

Pros: Minimal processing and transportation, little or no end-of-life costs (it is composted or dispersed), and rustic aesthetic appeal are some of the advantages.

Cons: Growing, harvesting and processing vegetation for landscape materials will increase labor costs, which is the greatest and most contentious expense in landscape maintenance. The work also requires workers with horticultural knowledge. Additionally, many of the materials below are short-lived and require frequent replacement.

Nomenclature

Branch: A limb that grows from a trunk of a plant. Shrubs without a trunk will have several or many main branches. Used for borders, handrails, stakes and small walls.

Branch Ridge and Branch Collar: The union where a branch meets a trunk or a stem meets a branch. While oftentimes bulbous, a branch collar does not necessarily bulge. Never cut through this section, always cut just above, as a collar/ridge injury can be fatal to a plant.

Coppicing: A pruning technique that requires cutting a tree nearly to the ground, inspiring the stump to sprout with a profusion of new shoots. These shoots are then harvested at various intervals for a variety of uses. Some of the uses include charcoal, edging, fences, fiber, firewood, formwork, furniture, poles, posts, thatching, stakes, trim and weaving.

Companion Plants: Companion planting is using an assortment of plants to create better outcomes for a primary group of plants. This primary group might be an orchard, residential shade garden, or community food garden. Companion plants are used to help with fertilization, pest control, and/or pollination.

Crown: There are two definitions relating to horticulture. The first is where the trunk or main branches meet the roots. While oftentimes bulbous, the root crown does not always bulge. The second is the size and shape of a tree's canopy.

Crown-Sprout: The new growth that sprouts after a plant has been cut at its crown. The sprout can be above the crown, called watersprouts, or below the crown, called root shoots and suckers. The intentional cutting of a plant for harvest is called coppicing.

Earthen: Inorganic materials such as clay, gravel, minerals, rocks and sand that are used to create adobe, blocks, buildings, foundations, plasters/renders, ponds, steps and walls.

Milled Product: Wood products made from the log of a tree (saw log) in a sawmill. See the chapter on Timber.

Non-Milled Product: Wood products that do not come from a sawmill, demand less processing, and are of lesser value. Some of the products are firewood, fiber, fuel, mulch, planks, poles, and pulpwood.

Pollarding: A pruning technique that frequently prunes back to the main trunk. Pollarding is used to keep a tree small or to produce products such as fiber, fodder, and woody products. Pollarding is a better option for wood generation than coppicing in landscapes with grazing animals because it keeps the foliage above their mouths.

Propagules: Any part of the plant that can be used for propagation. Propagation can be asexual, as in the case of cuttings and divisions, or sexual, as in the case of seeds and spores.

Root: The underground portion of a plant. Roots are utilized for a variety of purposes, such as dye, fiber, food and propagules. Aerial roots, from such trees as the rubber tree (*Ficus elastica*), are highly flexible and can be used for cordage, borders, and weaving.

Root Shoot: Shoots that grow from below the crown or along the roots from adventitious buds. Suckers are a root shoot. They are clones of the plant they sprout from. Harvested for propagation, poles, and weaving.

Snag: A dead, decaying and standing tree. Snags are good for habitat, but poor for most human use.

Stem: A limb growing from a branch.

Sucker: Vigorous new shoots that grow from the roots, most commonly from just below the crown. They may indicate stress or disturbance. Suckers are quick growing and generally straight. Because of their bendable nature, they are good for borders, weaving and some types of structures.

Timber: Wood that has been processed for building or carpentry. Refer to the Timber chapter for greater detail.

Trunk: The part of a tree that starts just above the crown and ends at the highest scaffolding branches. The trunk is sometimes referred to as the Useable Length.

Twig: A small limb growing from a stem. Used for carbon in compost, dye, and as a fire starter.

Watersprout: Vigorous new shoots that grow above the crown and from dormant buds along the trunk and branches. They usually indicate stress. Watersprouts are weakly attached and quick growing. Because of their bendable nature, they are good for borders and weaving.

Wattle: The interweaving of light and flexible material to control erosion and/or make fences, roofing structures and screens.

Landscape Materials that Can be Generated Onsite

Earthen Products

So much goodness can be pulled from the ground. Buildings and landscape features can be constructed using inorganic materials harvested onsite. People have been doing so for thousands of years.

While some earthen materials can be found on the surface, such as gravel, rocks and sand, some things are found in the subsoil, such as clays and minerals. Subsoil is sandwiched between the topsoil (and its organic matter) and the parent material or bedrock. A subsoil is where the clays and mineral deposits are found and is between 8 inches and 3 feet deep.[1]

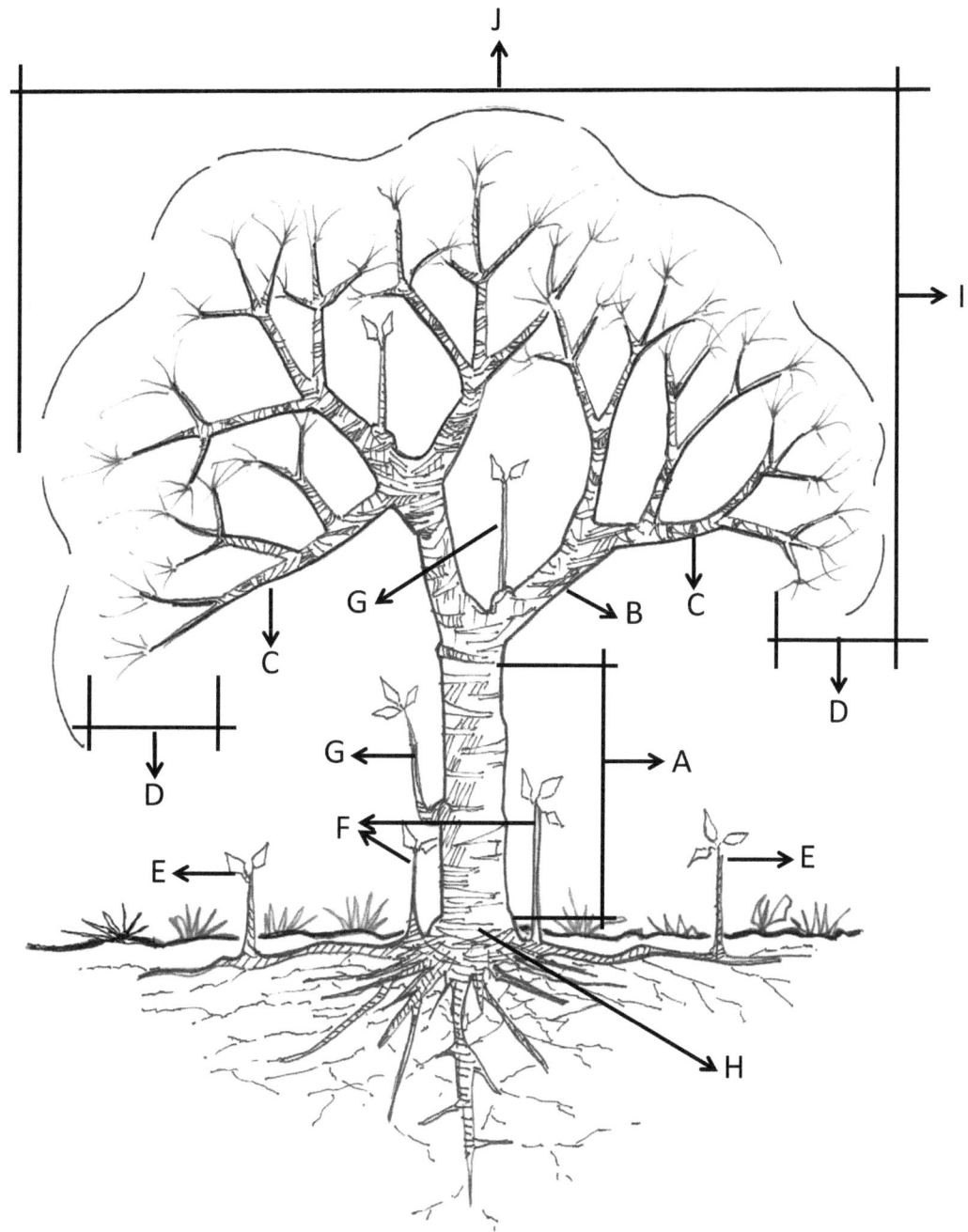

Figure 5.2   **This illustration highlights some of the nomenclature used to describe the parts of a tree.**

A. Trunk B. Branch C. Stem D. Twig E. Root shoot F. Sucker G. Watersprout H. Root crown I. Leaf crown J. Canopy

OVERVIEW

Uses: Earthen materials are used to create adobe, binder (cement), blocks, buildings, foundations, plasters/renders, ponds, steps, and walls.

Note: Adobe is historically translated to meaning mudbrick. However, in modern times it also refers to a type of architecture.

Pros: Earthen materials are recyclable. They require little energy or expense to make them useable. They can be harvested onsite. They are non-toxic and almost non-polluting (dust and particulates). And the buildings these materials create are healthy for the occupants and possess good thermal regulation. Their ERoEIs can be fantastic.[2]

Cons: The excavation, processing and construction are labor intense, which dramatically increases costs. Also, few designers, engineers or contractors are well versed in earthen construction, which increases costs. Buildings with earthen material may run afoul of local building codes and governments because the nation has embraced mechanized methods of construction. Earthen features also require more maintenance than industrial made products.

Figure 5.3   **The bulk of this building was made from Earth just 500 feet away. The structure was built using Earth bags and the stucco is called Super Adobe. The depression this structure created is used as an infiltration basin for a large housing and parking complex. Picture taken at Uncommon Good in Claremont, CA.**

Fertilizers

Fertilizer is a concentration of nutrients that boost plant growth. They are either liquid or solid. Fertilizer is either grown in place and worked into the soil or grown and then processed into a nutrient dense medium (liquid or solid). There are three broad types of fertilizers that can be grown: in place crops, humus crops, and tea crops.

IN-PLACE CROPS

Whether called cover crop, green fertilizer, green manure or nurse crop, the plants below are the ones o that are grown in place and then worked into the soil, where it slowly decomposes, leaving valuable nutrients within the top 1 foot of soil. In-place fertilizers are great for nutrient demanding landscapes, such as annual food production. These plants are divided between nitrogen fixers and biomass contributors.

Nitrogen Fixing Crops: Nitrogen fixers synthesize nitrogen from the atmosphere with the help of bacteria. These plants, with the help of bacteria, provide the nutrient most needed by plants. Many of the best nitrogen fixers come from the legume family (*Fabaceae*) and include annuals such as alfalfa, bean, pea and vetch; perennials such as clover, lupine, trefoil and wild licorice (*Glycyrrhiza*); shrubs such as broom (*Cytisus* and *Genista*), *Ceanothus*, false indigo (*Amorpha*), mountain mahogany (*Cercocarpus*), myrtle and wild indigo (*Baptisia*); and trees, such as *Acacia*, alder, locust, mesquite, and redbud.

Biomass Crops: Biomass contributing plants break up soil, pull nutrients from deep in the soil, suppress weeds, and add a lot of carbon, all of which improves a soil's tilth, increases numbers of microorganisms, helps the exchange of gasses, and improves soil's ability to hold onto water. Some of the best plants are barley, buckwheat, cheeseweed, millet, mustard, oat, radish, rapeseed, rye and wheat.

HUMUS CROPS

Humus is called black gold. Bacteria, bugs and fungi digest piles of organic matter into the elements that plants can readily synthesize. Humus is highly decomposed and nutrient rich. While any plant can be composted, the best, and easiest to grow are the fleshy pioneer species, the first to colonize after a disturbance. Some of the best pioneer species are annual grasses, borage, broom, cheeseweed, clover, comfrey, dandelion, fennel, filaree, fireweed, medick, mustard, nettle, radish, and *Rumex*. Also see Mulch and Compost further below.

TEA CROPS

Fertilizer teas are made from compost, humus, and/or manure which is then brewed as a liquid. So, in effect, it is a concentrate of a concentrate and hence power packed. Its nutrients are immediately available and digestible. They are often used as a quick remedy. Any of the plants previously mentioned can be composted for teas. Coffee grounds and manures are often used as well.

FERTILITY OF URBAN SOILS

Urban soils generally have more nutrients than native or natural soils. Research shows urban soils have higher amounts of potassium and phosphorus, and in some cases, nitrogen, the nutrient most needed by plants.[3] These macronutrients come from decades of gardeners supplying compost and fertilizers, of dead material left in the soil to rot, of pets using gardens like bathrooms, and of nitrogen deposition from tailpipes. Said another way, we contribute more than we extract.

One of the best ways to identify rich urban soils is to look for a bulge. Any soil that has swelled above its original grade is probably rich with carbon, gases, and microorganisms. As a rule: Do not fertilize these bulging soils unless a nutrient deficiency becomes apparent.

Figure 5.4   This landscape and parkway are 12 inches above the original grade in some places. The sidewalk is the lowest feature. None of the plants in this picture require fertilization because of the rich and bulging soil.

Fibers: Cordage (rope and string) and Weaving

Whether thread, twine or rope, baskets, mats or screens, there are many plants that produce the ideal fibers for cordage and weaving. Refer to the chapter on Craft and Textiles for more detail.

Mulch, Compost, Humus

Mulch is any a material laid over the top of soil to satisfy a goal. There are three broad types of organic mulch and five broad goals.

Notably, creating mulch, compost or humus should be the last use for woody refuse. The reason is because mulch is the last iteration of biomass before it turns back to its basic components, which are primarily atmospheric carbon, hydrogen and nitrogen. Generally, trunks, branches and stems make mulch; stems, twigs and leaves make compost; and twigs and leaves make humus.

TYPES OF MULCH

Mulch: The woodiest of the mulches. Used for water conservation, erosion control, and weed suppression. Weed suppressing mulches are carbonaceous and slow to decompose (see the Pesticide section below). They are sold as chippings, wood chips or bark shavings. Mulch will eventually decompose and turn to compost.

Compost: Nutrient providing, partially decomposed, and used as an amendment and top dressing, compost is good for soil protection, water conservation, and as a mild fertilizer. Good compost will be created from a pile that has a carbon to nitrogen ratio of 2:1 to 3:1,

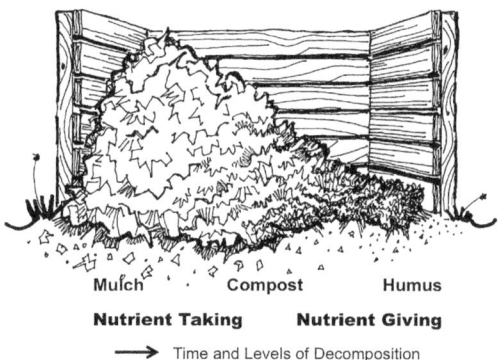

Figure 5.5   **This illustration shows the impact of time and decomposition on woody material. It starts as mulch, turns into compost, and eventually becomes humus when it is extensively decomposed.**

meaning 50% to 67% of the pile is carbon dense material, like branches, and 50% to 33% is nitrogen dense leaves and fleshy material.

Humus: Compost that is nearly, if not totally decomposed. It is worked into soil and has fantastic water holding and nutrient providing qualities. Good humus, however, requires time and effort. See the section above on Fertilizers.

FIVE BROAD USES FOR MULCH

Conservation: Mulch is laid over the soil to reduce surface temperatures, ultimately reducing water loss from evaporation and transpiration. It is an excellent strategy in dry and hot areas with little voluntary ground cover. Any woody mulch will work.

Decorative: Mulch that is used when it is not needed. Some of the most beautiful mulches include cedar, coco hulls, redwood, and softwood sawdust.

Erosion Control: The mulches used to slow topsoil loss are heavy, woody, slow to decompose, have a lot of surface area, and adhere and mat with other materials. Some of the best plants for erosion controlling mulches are bark shavings that come from cedar, cypress, *Eucalyptus*, juniper, pine, and willow.

Fertilizer: Plants grown to either be tilled directly into the ground or added to a compost pile. Please refer to the Fertilizer section above.

Weed Suppression: Mulch helps suppress weeds in three ways. It smothers the soil and deters seeds from sprouting, it blankets the soil and prevents weed seeds from being able to reach the soil; and it exudes chemicals that inhibit seeds from growing, such as *Eucalyptus* (see below for greater detail). No less than 4 inches of dense material is needed for effective weed suppression.

Figure 5.6   **Humus is often called black gold because of its powerful impact on plant health. Not only does humus contain vital nutrients, but it is loaded with beneficial microorganisms and acids as well.**

Pest Control[4]

Given the ecological and human health impacts from the widespread use of industrial pesticides, conscience gardeners need alternatives. Luckily, there are many healthy options, many of which can be grown.

HERBICIDE ALTERNATIVES

There are three primary crops for weed suppression: trees, mulches, and aggressive plants.

Weed Suppressing Trees: Many trees are highly competitive and will use chemical and smothering tactics to deter, suppress and/or kill competing plants. These trees will reduce weeds, which makes them a great plant for large properties, such as college campuses, industrial parks, and utility easements.

Some of the trees that are known to suppress weeds include *Acacia*, black walnut, elderberry, *Eucalyptus*, juniper, mesquite, pine, redwood, and tree of heaven.

Mulch: The best weed suppressing mulches have certain characteristics. They will be high in carbon (no leafy matter), high in oils, and chipped from wood that is slow to breakdown. Additionally, the greener (newer) the chipped material, the more effective it is.

Figure 5.7   **Pine needles do a wonderful job of suppressing weeds. The needles falling from the trees pictured are spread across the soil instead of being hauled away as greenwaste.**

The most successful mulches for weed suppression are recently chipped *Acacia*, camphor, cedar, *Eucalyptus*, juniper, oak, pine, and *Pittosporum*. Large, thick mulches are preferred over fine, thin mulches. No less than 4 inches of mulch is needed for suppression.

Aggressive Ground Covers: The best defense is often a strong offense. Using plants that can outcompete weeds is a great time-saver. The plants most likely to beat weeds share some general characteristics:

- They trail and root across the top of the soil.
- They are self-repairing and spring back from injury.
- Their foliage blocks the sun from striking the soil.
- They may have aggressive roots near the surface that hog water and nutrients.
- They may be prolific seeders.

INSECTICIDE ALTERNATIVES

Plants have been grown and used for pest control for thousands of years. Some of the best plants are highlighted below. These plants are organized into four categories: Plants that attract beneficial insects, plants that repel pests, plants that trap unwanted insects, and plants that are used for concoctions.

Figure 5.8   The Algerian ivy (*Hedera canariensis*) pictured here has been holding this slope and suppressing weeds for over 30 years. It has done so with little water, fertilizers, and weeding. Aggressive ground covers are ideal in urban areas where they cannot escape to wild areas. Picture taken at Orange Coast College, Costa Mesa, CA.

Figure 5.9 **Shown above are the four aspects of alternative to insecticides: Attracting predatory insects, repelling unwanted insects, trapping insects, and creating concoctions to either kill or repel unwanted insects.**

Plants that Attract

The purpose of these plants is to attract beneficial predators—the insects that will eat the pests. These plants do this by providing food, water and/or breeding and resting opportunities. Plants that attract beneficial insects include alyssum, angelica, aster, basil, blackeyed Susan, blanket flowers, buckwheat, clover, coreopsis, coriander, cosmos, dill, fennel, feverfew, mustard, sage, tansy, tidy tips, and yarrow.

Plants that Repel

These are the plants that some pests actively dislike and try to avoid. They include angelica, basil, bay, camphor, castor bean, catmint, chive, clover, daffodil, elderberry, *Eucalyptus*, garlic, horehound, juniper, leek, marigold, mint, mosquito plant, *Myoporum*, mustard, onion, petunia, rosemary, sagebrush, southernwood, tansy, and tobacco.

Plants that Trap

The goal of these plants is to attract pests, pulling them away from favored plants. Though they are sacrificial, they still need attention—the pests they attract will become a problem if not handled. Pruning or removing the infested vegetation regularly and then replanting is necessary. Some of the plants that trap unwanted pests include basil, chervil, clover, *Datura*, fennel, lambsquarter, marigold, nasturtium, Pelargonium, sorrel, sunflower and wild radish.

There are three strategies that will make the plants discussed above more impactful. First, ensure that something is in bloom most of the year: Spring, summer and fall flowers are needed. Second, pick at least three plants for each season, meaning that at a minimum a landscape will have nine types of companion plants. Third, plant the same plant in groups of twos and threes: The greater the mass, the greater its effect.

Figure 5.10 **Not only is nasturtium (*Tropaeolum majus*) deliciously edible, but it also helps protect food and foundational plants. It does this by attracting the plant damaging insects, allowing the gardener to cut away the infected areas and toss the pests away.**

Plants for Specific Concoctions

These are the plants grown and then processed for specific applications. The types of pest-controlling applications include powders, sprays, and teas.

Ants: Lemon and spearmint makes teas and sprays.

Aphids: Garlic, lemon, mugwort, onion, shallot and tobacco make teas and sprays.

Cats: Black pepper, Cayenne pepper and mustard make powders.

Deer: Garlic, capsaicin, peppermint, and rotting eggs are spread around threatened plants.

Fleas: California bay and *Eucalyptus* mulch around doghouses, kennels and houses, and in nylon stockings hung inside a house.

Grasshoppers: Hot peppers and onions make teas and sprays.

Paper Wasps and Yellow Jackets: Lemon and mint teas and sprays used where wasps and hornets congregate.

Mites: Capsaicin is distilled for a spray.

Gophers and Moles: Blackberry, castorbean, elderberry and hot peppers are shoved holes and runs.

Mosquitos: Citronella grass (*Cymbopogon nardus*), lemon-scented gum (*Corymbia citriodora*), mosquito plant (*Pelargonium citrosum*), and tea tree (*Melaleuca alternifolia*) are distilled for oils that are used in candles, lotions, cleansing sticks, and sprays. Catnip, lemon balm, ngaio tree (*Myoporum laetum*), mugwort (*Artemisia douglasiana*), rosemary leaves, and white sage (*Salvia apiana*) are rubbed on clothes and/or distilled into a tea and sprayed on clothes and skin. And sagebrush and wormwood (*Artemisia californica*, *A. tridentataI* and *A. absinthium*) are burned.

Thrips and whiteflies: Mugwort and tobacco teas mixed with oil and water to make a spray.

Propagules (new plants)

Propagating plants instead of buying them is one of the most underutilized, yet beneficial aspects of urban harvest. It saves car trips, packaging, and plastics. The returns on energy are fantastic. The source from new plants is not only your landscape, but also—and with permission—your neighbors' and community's plants, more so if the propagule grows into public areas.

There are drawbacks to propagating a site's vegetation and they are the reasons why people, organizations and businesses do not do it. Propagating can consume a lot of space. It may require specialized equipment and skill. It requires time. And propagating forces the land manager to plan months, if not years in advance.

The subject of plant propagation is vast. Highlighted below are the types of propagation, not the specific practices.

FOUR MOST COMMON TYPES OF PROPAGULES

Cuttings: Some leaves, stems and roots can be cut and regenerated to produce new plants. This is a form of asexual propagation and produces a clone of the parent. Perennials, shrubs and trees are commonly propagated by cuttings. Below are some of the plants that can be started by cuttings:

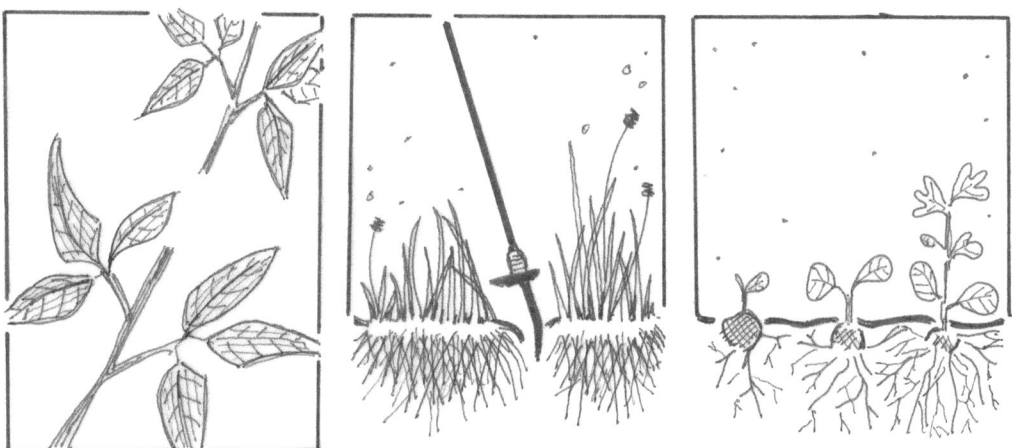

Figure 5.11   **Shown above are the three most common ways to propagate a plant: Cutting, division, and seeding.**

- Growing tip and/or stem: Avocado, bay, blackberry, carob, *Ceanothus*, citrus, dragon fruit, elderberry, grape, mulberry, passion fruit, redwood, rose, sumac, and *Salvia*.
- Root: Blackberry, fern, *Ficus*, *Forsythia*, honeysuckle, ivy, *Liquidambar*, mint, oregano, potato, peppertree, *Prunus*, raspberries, redwood, rosemary, strawberry, willow, and yarrow.
- Leaves: African violet, cape primrose, *Gloxinia*, *Plectranthus*, *Sansevieria*, some begonia species, and many succulents, such as *Crassula*, dragon fruit, *Kalanchoe*, *Opuntia* and *Sedum*.

Grafting is a form of cutting and marries two plants together. Typically, the union includes a strong rot resistant rootstock married to a productive canopy; both the rootstock and canopy coming from a scion (a parent plant).

Divisions: Clumping and pupping plants can be divided by either pulling them or cutting them apart. Some of the most common plants that are divided include *Agave*, *Aloe*, banana, cattail, coral bells, daylily, fern, ginger, *Iris*, reed, rhubarb, strawberry, sugarcane, yarrow, and most perennial clumping grasses, such as *Carex* and rye.

Seeds: Starting plants from seeds is sexual and the union of an egg and pollen. It creates a new individual. Most annuals are started by seed. While seeds are generally easy to

Figure 5.12   **These homeowners use their property, and the plants on it, to enrich their neighborhood. They sell plants that they started from their landscape's plants. This is not a money-making venture, but a community building venture. Their herbs and succulents are propagated from cuttings and divisions.**

germinate, some need some type of treatment to break their dormancy, which might be pH bath, charring, freezing, scarification and/or submersion.

Spores: Unless you are aiming for more algae, fungi or lichen, using spores for propagation is generally restricted to ferns, and even then, it is a hobby as ferns are easy to propagate from divisions and root cuttings.

Screens

Screens can either be grown in place or constructed.

GROWN IN PLACE

As any nurseryperson will tell you, buying plants to create a screen is one of the leading reasons for buying outdoor plants. This book makes many recommendations for grown screens, such as those that produce food (Food chapter) and remove gaseous and particulate pollution (Public Health chapter).

GROWN AND CONSTRUCTED SCREENS

Plants used to make a screen do not have to be structurally sound, but they should be abundant and easy to work with. The most common species to plant, harvest or scavenge for screening material includes *Arundo*, bamboo, cattail, fan palm, hemp, ivy, *Miscanthus*, palm fronds, planks, and sugar cane.

Woody Products

Woody crops are substantial, long lasting, and have many practical uses. Covered below are the plants that are harvested for borders, handrails, logs, planks, poles, stakes, surfaces, and wattle.

BORDERS/EDGING

What gardener doesn't need to keep people on paths, plants out of specific fields, and define areas of interest? Borders are essential to a safe and successful urban landscape. Fortunately, urban landscapes can easily produce these materials in abundance.

Edging material can be big and blunt, like trunks, or be small and sinuous, like branches and stems. While any trunk, branch or stem can be laid on the ground to create a

Figure 5.13 **Milling savaged lumber from cities across the Bay Area (CA), Richmond Greenwaste Recycling supplies borders, fencing, planks, posts, retaining walls, and steps to the cities of Oakland and San Francisco.**

border, the best wood will be hardwood and long lasting. If bending or weaving is required, always work with just-cut (greenwood) branches, stems and watersprouts.

Some of the common trees that produce material ideal for borders include *Acacia*, ash, birch, black walnut, camphor, cedar, cypress, fir, incense cedar, ironwood, juniper, *Liquidambar*, madrone, maple, pine, *Prunus*, oak, redwood, sycamore, and the trunk of a fallen tree.

HANDRAILS

Handrails demand a unique type of wood. The wood should be straight, amiable to drilling and polishing, and should not produce burrs and splinters as it ages. Some of the better shrubs and trees for handrails include bay (both California and sweet), cedar, empress tree, fir, ironwood, oak, *Prunus*, *Rhus*, sycamore, tanbark oak, and walnut.

LOGS

These are the branches and trunks greater than 1 foot in diameter. Logs are used for benches, steps and retaining walls. Logs artistically used in a landscape provide bold and

Figure 5.14 **Instead of cutting up fallen trees and either burning or mulching them, this large public park tries to repurpose them. Here they are using the trunks to provide seating and respite. Picture taken at the Magic Johnson Recreational Center, Compton, CA.**

earthy effects, although they can be difficult to work with and move. A log with a diameter of over 2 feet can weigh over a thousand pounds, depending on its length and amount of moisture.

While any log will work, the longest lasting will be hardwood and/or rot resistant. Some of the best logs come from ash, black walnut, cedar, cypress, larch, madrone, oak, *Prunus*, redwood, spruce and sycamore.

PLANKS

Planks are rough-sawn lengths of a trunk. The bark is often kept on the boards. Planks are thick and between 2 to 4 inches. Straight trunks a best for planks because curved trunks tend to split. Planks are used for benches, decoration, screens, tabletops, walls and walkways. Ideal qualities for planks include interesting colors and patterns, rot resistant, and long lasting. Some of the trees include camphor, cedar, *Liquidambar*, oak, redwood, sycamore and walnut.

Please, refer to the chapter on Timber for a deeper discussion.

POLES

Poles are used for signposts, structural support, tree stakes, and as a part of wattle. Poles are long, straight and strong. A good size is between 2 to 5 inches in diameter and no less than 6 feet long; 8 feet and more is better.

Some of the best plants for stakes are bamboo, birch, *Catalpa*, cedar, cherry, Chinese pistache, empress tree, hoja santa, madrone, pine, redwood, sycamore and willow. Poles often come from coppiced plants (refer to box further below). Notably, poles capable of bearing loads come from the trunks of small trees and require 4 to 8 years to grow, depending on type of tree and desired girth.

ROOFING

Using vegetative material for roofing is both ancient and common. Vegetation is prepared and positioned to sheet rainfall and provide insulation. The roofing process is called thatching. Leaves and branches are the primary thatching material. Plants used for roofing include palms, reeds, rushes, sedges, and straw.

STAKES

Stakes have many uses in a landscape, such as pinning borders and nailing wattles, securing steps, and fortifying walls. They are between ½" to1¼" in width and 8" to 3' long, and made of smooth-skinned hardwoods. Some of the best stake-producing trees are *Acacia*, ash, bay black walnut, carob, Chinese pistache, *Eucalyptus*, *Forsythia*, *Liquidambar*, locust, Manzanita, mesquite, oak, *Prunus,* and sycamore.

TRELLIS

Trellises are essential in urban areas. They support the vines that provide contemplation, cooling, food, habitat, refuge, and a visual retreat. Trellises are constructed from logs, poles and branches.

Figure 5.15   Some of the useful things poles can be crafted into include tree stakes, borders, handrails, privacy fencing, and signposts.

Figure 5.16   **The branches from a large tree created the material for this trellis. It supports two Lady Banks' roses (*Rosa banksiae*) and shades a meditative area. Picture taken at Moonwater Farm, Compton, CA.**

WALKING SURFACES

Wood has long been used as a walking surface, decking being the most notable. Decking sits above the soil and involves many types of milled products, such as posts, joists and decking. Planks and wood rounds sit directly on the soil. Wood rounds are like stepping-stones and are cross sections of a trunk or large branch. They are generally 2 to 4 inches thick.

Some of the longest-lasting trunk rounds come from ash, black walnut, cedar, cypress, juniper, larch, locust, madrone, oak, *Prunus*, redwood, spruce, and sycamore.

WATTLE

Wattles are knitted structures and used for bank revegetation, erosion control, and screens. They are considered short-lived, although that depends on the material. Making a wattle involves driving many poles and then weaving any number of materials between them. Wattle is green and flexible.

Good wattle will be long, slim, and supple. Ideal diameter is from ¼" to 2". Wattle is made from branches, reeds, roots, shoots, sprouts and stems. Some of the best plants include *Baccharis*, bamboo, broom, *Catalpa*, elderberry, empress tree, *Eucalyptus*, *Forsythia*, hazel, island mallow, *Prunus*, *Rhus*, willow, and any bramble or vine, although they degrade faster.

COPPICING CROPS

Many species of shrubs and trees produce new shoots when cut just above their crown. These shoots, which are often straight, narrow and flexible, are harvested for a variety of uses, such as craft, fences, poles, screens, stakes, walls, and wattle.

Some of the best species for coppicing include *Acacia*, *Albizia*, alder, ash, black locust, buckeye, California buckeye, *Cassia*, *Catalpa*, chestnut, coast redwood, elderberry, elm, empress tree, *Eucalyptus*, *Forsythia*, hazel, Japanese pagoda tree, mesquite, hoja santa, oak, olive, *Prunus*, redwood, *Rhus*, silktree, sycamore, tree of heaven, and willow.

Figure 5.17   The coppiced empress tree, *Paulownia tomentosa* pictured, has long been used as a coppiced crop and for poles posts, and stakes.

Resources

BOOKS

Campbell, Paul D. *Survival Skills of Native California.* Gibbs Smith. 2009.

Lyle, John Tillman. *Regenerative Design for Sustainable Development.* John Wiley & Sons, Inc. 1994.

Lengen, Johan van. *The Barefoot Architect: A Handbook for Green Building.* Shelter Publication, Bolinas, CA. 2008.

Sorvig, Kim and Thompson, William. *Sustainable Landscape Construction, Third Edition: A Guide to Green Building Outdoors.* Island Press. 2018.

Weismann, Adam and Bryce, Katy. *Clay and Lime Renders, Plasters and Paints: A How-To Guide to Using Natural Finishes.* Green Books. 2015.

Notes

1 Van Damme, Henn and Houben, Hugo. "Earth Concrete. Stabilization Revisited." *Cement and Concrete Research*, vol. 114, December 2018, pp. 90–102.

  https://www.sciencedirect.com/science/article/pii/S0008884616308365

2 Lengen, Johan van. *The Barefoot Architect: A Handbook for Green Building*, Shelter Publication, Bolinas, CA., 2008.

3 Pouyat, Richard V., et al. "Chemical, Physical, and Biological Characteristics of Urban Soils." *Urban Ecosystem Ecology*, USDA Forest Service, Northern Research Station, ch. 7, 2010, pp. 119–152. https://www.fs.usda.gov/research/treesearch/36426

  Valuing Chaparral: Ecological, Socio-Economic, and Management Perspectives. Ed. Underwood, Emma C.; Safford, Hugh D.; Molinari, Nicole A.; Keeley, Jon E. Springer Nature 2018, p. 159.

  Zigas, Eli. "Is City Soil Really More Toxic Than Rural Soil?" *SPUR*, September 14, 2011. https://www.spur.org/news/2011-09-14/city-soil-really-more-toxic-rural-soil

4 *Invasive Plants of California's Wildlands.* Ed. Bossard, Carla C. et. al. University of California Press, 2000.

  Kent, Douglas. *California Friendly: A Maintenance Guide for Landscapers, Gardeners and Land Managers.* Douglas Kent+Associates, 2017.

  California Master Gardener Handbook, 2nd ed. Ed. by Pittenger, Dennis R. University of California, Agriculture and Natural Resources, 2015.

  And, the extensive work of University of California, Agriculture and Natural Resources, Statewide Integrated Pest Management Program. www.ipm.ucdavis.edu

Chapter 6

Public Health

Urban landscapes should be designed to benefit public health. You might think that would be Landscape 101. Yet, far too often, other goals, such as landscaping for butterflies, receive most of the attention. Public health is a low priority, if a priority at all. This needs to change.

There is ample evidence that landscapes designed with public health in mind reduce respiratory problems and allergic reactions and decrease heart rate and blood pressure. Plus, an environment that feels good will encourage its use and compel people to drop their blue screens and spend more time outdoors. And, of course, being outdoors strengthens bones, tones muscles, and improves immunity.

This chapter examines how to approach this goal. Most of the measures that need to be taken will be defensive, such as reducing allergens and pesticides. But there are proactive measures that can be taken as well such as planting to reduce particulates and other air pollutants and recognizing the restorative and therapeutic potential of gardens.

Human health is a complicated, multi-layered issue. This chapter only addresses a small part of the topic. This theme, however, is an integral part of the message of this textbook, and it is woven into every chapter, most especially in Food, Self-Care and Thermal Comfort.

Quick Overview of Landscaping for Public Health

Uses: The goals of landscaping for public health are boosting the bone density, immunity and respiratory health of everyone who visits that landscape as well as the populations living downwind of it. Public health landscapes can—and should—also aim at boosting morale.

DOI: 10.4324/9781003369752-6

Figure 6.1   **Covered in this chapter are air pollution, allergens, heat, pesticides, night lighting, and emotional wellbeing.**

Costs: Low to high.

Difficulty: The concept is easy to grasp and design for, but there may be reluctance to embrace these measures because of the shift in resources and increase in labor costs.

Best Scale: Public health and wellbeing should be prioritized in landscapes where every segment of the population is expected/encouraged to visit. These landscapes include those in dense downtowns, public properties, such as courthouses and libraries, amusement parks, and large commercial endeavors, like shopping malls.

ERoEI: What is the cost of being able to breathe deeply? What is the cost of an extra day, month, or year of health? Theoretically, all the strategies below should pay for themselves economically and energetically. However, the savings might not come directly but through things like reduced cooling bills and increased property values.

Pros: Health for all—landscaping for public health helps improve the health of all living things. Maintenance workers and residents, sea life and wildlife will all improve as fossil fuel use is reduced, urban vegetation is increased, and organic remedies rise in use. Furthermore, landscaping for health is fun: Flowers and fragrances abound, birds sing and swoop, and the landscape is quiet and calming.

Cons: Reducing the use of fossil fuel driven machines and herbicides will increase landscape labor costs, the greatest and most contentious expense in a landscape. Additionally, other environmental goals might have to play a lesser role, such as water conservation.

GETTING DIRTY IS ESSENTIAL FOR HEALTH

Want a robust smile? Grow root crops and get dirty. Want protection from viruses? Care for your landscape with your labor. Want a long life? Dig into your garden and its many wonders. No other recreational pursuit compares with gardening when it comes to improving personal health.

Mood:[1] There is a bacterium in soil called *Mycobacterium vaccae* that acts as an antidepressant by increasing norepinephrine and serotonin levels. This bacterium not only makes people feel happier, but has also been shown to improve emotional health, vitality, and general cognitive function.

Immunity:[2] Gardening is a mild exercise that has been proven to enhance levels of vitamin D and feelings of wellbeing, while reducing blood pressure and levels of stress, all of which improve our immune system.

Longevity:[3] The evidence is well-documented and immense. Gardening will extend your life. Mild exercise, getting dirty, and sunshine are fantastic for a longer, healthier life.

Figure 6.2   Pictured are a group of volunteers preparing an area for a food garden. The organizers encouraged the volunteers to bring their children because they were aware of the health benefits of getting dirty.

Nomenclature

Allergen: Anything that causes an allergic reaction.

Animal Pollination: Plant pollination that requires animal assistance. Common animal pollinators include ants, bats, bees, birds, moths and wasps.

Biological Volatile Organic Compound (BVOC): A hydrocarbon (gas) produced from living organisms (mostly plants) that possesses a potential chemical energy. BVOCs undermine respiratory health. BVOCs can be produced by animals, plants, and microscopic life.

Complete Flower: A flower defined by having four parts: Sepals, petals, pistils, and stamens. Complete flowers are generally non-allergenic.

Emotional Wellbeing: Feeling at peace with no anxiety or panic.

Female Species: A plant that produces female flowers, which have a stigma, style, and ovary. Female plants produce fruit and seed.

Four-Stroke Engine: A combustion engine that has four phases (strokes) for every one operating cycle: Intake, compression, power, and exhaust. Four-stroke engines burn cleaner than two-stroke because the oil, the piston lubricant, is separated from the fuel.

Gaseous Pollution: Gasses mixed in the atmosphere that endanger biological and public health. Gasses are classified as Primary and Secondary. Primary gasses come from the source and include carbon monoxide (CO), nitrogen oxides (NO_X), sulfur oxides (SO_X), and volatile organic compounds (VOCs). Secondary gaseous pollutants are caused when the primary pollutants react with heat and/or sunlight and transform, creating pollutants such as nitrogen dioxide (NO_2), sulfuric acid (H_2SO_4), nitric acid (HNO_3), ozone (O_3), and smog.

Integrated Pest Management (IPM): Managing horticultural pests in a manner that reduces the direct and indirect impacts of control. The subject is voluminous, but revolves around five principles: Identification, cultural practices, biological means, mechanical methods, physical remedies, and chemical controls.

Male Species: A plant that produces male flowers, which have an anther and filament. Male plants produce pollen, but not fruit or seed.

Particulate Pollution (PM): Solids and moisture mixed and suspended in the atmosphere. Particulate matter can be visible, such as ash in the sky, or microscopic. Elevated levels of particulate matter cause cancer, reduce lung capacity, and undermine general health and wellbeing.

Pesticides: A substance that disrupts, kills, or repels a living entity. The most common pesticides are herbicides, insecticides, fungicides, and rodenticides.

Respiratory Health: Unobstructed airway and full lung capacity define respiratory health.

Restorative and Therapeutic Landscapes: Landscapes that produce feelings of wellbeing. Restorative landscapes sooth everybody. Therapeutic landscapes help someone overcome a disability, such as a loss of hearing or sight.

Two-Stroke Engine: A combustion engine that has two phases (strokes) for every one operating cycle: Up and down. While two-stroke engines have greater weight to power ratio than four-stroke, they release more pollutants because the fuel is mixed with oil to lubricate the pistons.

Volatile Organic Compound (VOC): A hydrocarbon (gas) produced via industrial processes, paints and solvents, and landfills. VOCs undermine respiratory health. Also see Biological Volatile Organic Compound (BVOC).

Wind Pollination: Plant pollination that requires air movement to disperse pollen for sexual reproduction. The pollen from wind pollinated plants is typically allergic.

Air Pollution Reduction

The quality of air affects everyone and everything. It leads to an increase in cancer and a decrease in lung capacity, an increase in invasive species and a decrease in native species. Dense urban areas require landscape practices that can help dilute, filtrate, or mitigate air borne hazards.

Air pollution can be divided into two broad classifications: Gaseous and particulate.

- **Gaseous** pollution is then divided between primary and secondary. Primary gas pollution includes carbon dioxide and monoxide, nitrous and sulfur dioxides, and volatile organic compounds (VOCs). Secondary pollutants are the result of the primary pollutants reacting to other chemicals, heat, moisture and/or sunlight. Secondary pollutants include ground level ozone (O_3) and smog.
- **Particulate** pollution tears and scars our respiratory system. The number of fine particulates in an environment has been directly linked to life expectancy.[4] Most commonly we create particulates when we burn things, such as candles, cigarettes, firewood and mostly, fossil fuels. Dust, too, is a human byproduct and includes all the junk elevated while blowing, driving, grinding, sanding, and sweeping.

PRIMARY AND SECONDARY AIR POLLUTANTS[5]

Naturally, the first-out-of-the-pipes emissions peak over their sources. Traveling downwind these gasses react to sunlight and heat to form the secondary pollutants, which are even more harmful. Ozone and smog are the secondary byproducts of most concern. Ozone causes asthma attacks, increases hospital visits, reduces lung capacity, and contributes to premature death. These gasses can travel up to 100 miles beyond their sources.

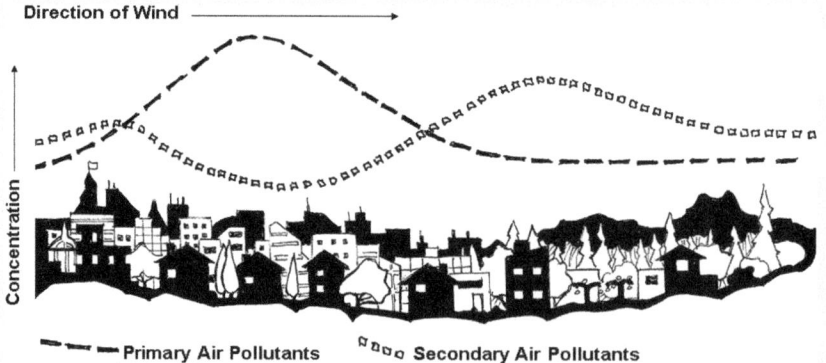

Figure 6.3   **Primary pollutants spike over their source and secondary pollutants spike further downwind, as this drawing illustrates.**

Design and Maintenance Strategies

The information below is divided between design strategies and maintenance strategies.

DESIGN STRATEGIES

Below are the design characteristics that help improve air quality. The strategies include cooling, dilution (airflow), managing BVOCS, and the removal of gaseous and particulate pollutants. The quickest way to reduce impact is to reduce the use of machines. Landscape equipment, such as tractors, rototillers, mowers, and blowers, are notorious for the air pollution they produce.

Cooling a Landscape

Urban areas are naturally warmer than wild areas and due to climate change these areas are getting even warmer. The impacts of increased heat are wide ranging—affecting everything from public and ecosystem health to coastal erosion and wildfire frequency. Cooling our communities is vital. Please, refer to the chapter on Thermal Comfort for far greater detail on cooling strategies.

Increasing Wind

Air pollutants will naturally settle and accumulate in areas with little or no wind. Increasing the speed of wind mixes and dilutes the amount of gaseous and particulates per gulp of air, extreme weather discounted. However, dilution is a strategy for personal health, not public health. Increasing airflow only transports the pollutants elsewhere. If a community wants an overall reduction of air pollutants, then they would need to trap pollutants by slowing airflow in areas with few people and speed airflow in areas with many people.[6]

Figure 6.4   **The illustration above shows that particulate pollution will increase immediately around an obstruction but provide an overall reduction when the wind picks back up to speed.**

Managing BVOCs

Volatile organic compounds (VOCs) are organic pollutants that impact air quality and human health. VOCs react to nitrogen oxide and sunlight and form ozone and smog. VOCs come from a variety of landscape practices, such as engine use, paints, sealers, and plants.

Many types of plants produce oils and resins to conserve water, repel insects, suppress competition, and/or attract pollinators. When these plants breathe, they release biogenic volatile organic compounds, BVOCs. The amount of BVOCs plants produce increase as temperatures rise. In some cases, BVOCs can be as much as 10% of all VOCs emitted in an urban area.[7]

Low BVOC producing plants are preferred in both hot environments and places where people are sensitive to air quality, such as hospitals, retirement communities, and heavily populated public areas.

Pollutant Removal[9]

Urban vegetation can improve air quality. Trees and dense vegetation will remove airborne particulates, BVOCs, ozone, and nitrogen and sulfur dioxides. The percentage of air pollution removal varies from .09% to 13%. The amount of removal is related to plant density, leaf surface area, humidity, temperature, and wind speed.

Particulate Removal

Is there dust on a plant? If so, particulates have settled out of the air. While all plants can slow airflow and screen particulates, some plants are better than others.

Cypresses and junipers, with their dense and intricate and immense leaf surface areas, are particularly good for slowing and screening. Other plants, such as silver birch (*Betula pendula*), yew (*Taxus baccata*), and elderberry (*Sambucus* spp.), pull particulates out with magnetic qualities.[10]

Table 6.1   High BVOC Trees[8]

Callistemon spp.	Bottlebrush
Casuarina spp.	Australian Pine Beefwood
Cupaniopsis anacardioides	Carrot Wood
Eucalyptus spp.	Gum
Ficus spp.	Fig
Koelreuteria spp.	Goldenrain Tree
Liquidambar spp.	Sweetgum
Myrica spp.	Myrtle
Nyssa sylvatica	Black Gum
Picea spp.	Spruce
Platanus spp.	Plane Tree/Sycamore
Populus spp.	Poplar/Cottonwood
Quercus spp.	Oak
Rhamnus spp.	Many Varieties
Robinia spp.	Locust
Salix spp.	Willow

Figure 6.5   **With its intricate, dense and sticky scale-like leaves, junipers are effective at trapping particulates. Some of the best landscapes for junipers are where people and cars coexist, such as busy residential streets and parking lots.**

Ideally, low growing junipers should be planted in street mediums and parkways to capture the plastics, metals, rubber and soot that flies from cars. Taller conifers, shrubs and trees should be planted as close to roads as possible without obscuring views.

Gaseous Removal

According to researcher Morikawa Takahashi urban landscape designers should be recommending vegetation that benefits from pollution rather than vegetation that is resistant to pollution.[11] Some of the gaseous pollutants, particularly the nitrogen oxides, stimulate plant growth. Plants that reduce gaseous pollution are best in areas where people congregate, such as courtyards, entryways, and the entrances of commercial and municipal buildings.

Useful Plants that Utilize Gaseous Pollution

Big leaves and fast growth define a plant that gulps carbon and nitrogen oxides. Of the hundreds of plants that fit this description, below are a few:

- **Annuals**: Corn and sunflower.
- **Big Grasses**: Bamboo and sugarcane.
- **Vines**: Blackberry, grape and passionflower.

- **Trees**: Avocado, carob and loquat.
- **Weeds**: Black mustard, castorbean and fennel.

MAINTENANCE STRATEGIES/SOURCE REDUCTION

The way we manage our landscapes has a large impact on local air quality. We can either increase the number of pollutants or reduce them. Below are maintenance strategies aimed at source reduction and the management of particulate pollutants.

Restrict Blowers

Blowers have three health impacts on humans. First, they shove chemicals, dust, fecal matter, mold spores, particulates, pesticides, pollen, and toxins back into the atmosphere. Blowers greatly elevate particulate pollution. Second, they produce hydrocarbons and

Figure 6.6   **The plants in this picture serve three goals. They dramatically cool the area, they gulp gaseous pollution, and they provide food. Pictured are hoja santa, loquat, mango, passionflower, and spearmint. Picture taken at the Lyle Center for Regenerative Studies, Pomona, CA**

oxides (see below). Third, blowers are loud. They typically run at 95 decibels and anything over 85 decibels is considered harmful.

Blowers should be avoided or reduced in low wind environments and around vulnerable populations. Places where the use of blowers should be restricted include convalescent facilities, dense downtowns, government buildings, hospitals, and large apartment complexes.

Restrict Engine Use

Because of its small and portable nature, fossil fuel driven landscape equipment is notoriously polluting. Carbon monoxide, nitrogen and sulfur oxides, particulate matter and VOCs are byproducts of landscape equipment, particularly two-stroke engines.

The ideal strategy is to reduce the use of fossil fuel powered machines. Brooms would be used instead of blowers, loppers and saws instead of chainsaws, and machetes and sickles instead of weed whackers.

If the use of machines cannot be avoided, then upgrade to an electric or four-stroke machine. No matter the type of machine, performing regular maintenance on the machine will improve efficiency and reduce its number of air-borne pollutants.

Protect Mature Vegetation

Large plants are much more effective at pollutant removal than small plants—mass equals mitigation. Assaults on mature vegetation can take many forms, such as construction damage, poor pruning practices, damage from pedestrians and cars, earth moving, and improper irrigation. Healthy and large vegetation rewards our protection with cleaner air.

Manage Particulates

Plants do not necessarily remove particulates from the air. Rather they grab and store them. Unless something is done, these particulates can be propelled back into the atmosphere by jarring the plant or high winds. The goal is to get the particulates to settle into the soil, where they can be biologically assimilated, synthesized or transformed, and this means washing the plant.

The plants within areas of poor air quality should be hosed down once a month during the hotter times of year. This hosing will do two things. First, the water will grab the particulates and send them to the soil. And second, the water will wash the leaves, making it easier for the plant to breathe, photosynthesize, and defend itself against pests.

Transportation

Moving people and materials in and out of managed landscapes is not only the greatest use of energy (and source of air-borne pollutants) in a landscape, but also the leading cause of death within the green industry. Transportation is costly.

Maintenance costs are a product of the landscape design. A native landscape may need professional attention no more than once every three weeks; a turf landscape requires weekly maintenance with a suite of machines. Besides design, there are other strategies to reduce transportation costs, such as consolidating trips (good planning), shopping locally, carpooling, avoiding rush hour traffic, and doing more of the work with the people on site. Importantly, vehicles must be regularly maintained to ensure optimum performance.

Cooling Urban Landscapes

Our warming planet gravely impacts the health of people and wildlife. Warming is linked to the formation of harmful air pollutants, like low-level ozone and smog. It increases the chances of heat-related illnesses, such as heat strokes. It increases the temperature of runoff, contributing to the degradation of waterbodies. And it impacts landscapes many miles away.[12]

There are many strategies for cooling landscapes naturally. Please, refer to the chapter on Thermal Comfort for greater detail.

Reducing Allergens

Over 50 million Americans suffer from allergies. Symptoms range from runny eyes and nose to severe headache and constricted breathing. One of the most widespread symptoms is itching.

Many of our allergens come from urban activities and landscapes are known contributors. Landscapes can be a source of BVOCs, ozone, particulates, pesticides, pollen, and rubber, all of which makes many people miserable. Pollen is one of the more potent allergens. Fortunately, it is controllable in managed/urban landscapes.

Figure 6.7   **Shown above are the primary strategies for reducing allergens: avoid wind-pollinated plants, restrict the use of blowers, and mow or prune vegetation before it can flower and produce pollen and allergens.**

Design and Maintenance Strategies

Because BVOCs, ozone, particulates and pesticides are more than just allergens, they are covered separately and elsewhere. The design and maintenance recommendations that follow are aimed solely at reducing the amount of allergenic plant pollen.

DESIGN STRATEGIES

The pollen from wind-pollinated plants is one of the worst allergens in urban areas—so much so that it is tracked in newspapers and on television. Wind pollinated plants should be avoided in dense areas and/or near vulnerable populations, such as the elderly and frail.

There are two types of wind pollinated plants. The first are the plants that can pollinate themselves. Called monoecious, this type of plant has both female and male parts. Fig is a good example. The second type of plants are those where each species has a sex, just like us. They are called dioecious. Male plants produce the pollen and allergens and the females produce the fruits and seeds. Sycamore is a good example. Unfortunately, most cities and homeowners avoid female plants because of their mess, unintentionally increasing allergens.

Figure 6.8   **Pictured is a community center in one of Los Angeles County's most polluted and warm communities: South LA. Every plant in this picture produces an allergen and every tree BVOCs. The goals for this design were native-ness, not public health, despite being surrounded by thousands of people that need reasons to get outside, not stay in. Picture taken at the Magic Johnson Recreational Park.**

Reducing levels of pollen in urban areas means choosing animal pollinated plants. Whether pollinated by ants or bees, moths or birds, a majority of animal pollinated plants are not allergenic. The flowers that attract animals are typically large, showy and slightly fragrant. Notably, not everything fits neatly into categories. Pine trees are a good example. They are not pollinated by animals and produce copious amounts of pollen, but because it is coated in wax it is not a severe allergen. And, though most animal pollinated plants don't contribute to the problem, there are some that do.

Plants to Avoid in Around Vulnerable Populations

There are hundreds of allergenic plants. These offenders come from every category: Annuals, perennials, grasses, shrubs, and trees. The emphasis of this list is on the persuasive pollen producers commonly used in ornamental landscapes. There are also many animal-pollinated plants that are allergenic, and they are also listed below.[13]

Table 6.2   Wind Pollinated Allergenic Plants

Common Name	Botanic Name	Notes
All grasses, including barely, bent, Bermuda, blue, brome, fescue, fountain, redtop, rye and velvet.	Genus includes *Hordeum, Agrostis, Poa, Bromus, Festuca, Pennisetum, Lolium,* and *Holcus.*	These are used as turf, erosion control, pasture or ornamentally. Always mow just before flower production.
Alder	*Alnus* spp.	Riparian native.
Artemisia—many names	*Artemisia* spp.	Relative of ragweed. The odor can trigger reactions in some people.
Ash: Male	*Fraxinus* spp.	Males are the variety most available.
Chinese Tallow Tree: Male	*Triadica* spp.	Street tree whose sap is also poisonous.
Coyote Brush: Male	*Baccharis pilularis*	Plants are also poisonous.
Cypress	*Cupressus* spp.	Massive amounts of pollen for up to 6 months a year.
False Arborvitae	*Thujopsis dolabrata*	A stately conifer that is good at catching particulates.
Fern Pine, Yew Pine: Male	*Podocarpus* spp.	Widely used as a screen or street tree.
Hickory, Pecan	*Carya* spp.	Popular fruit-bearing trees.
Juniper: Male	*Juniperus* spp.	Females, the ones to get, have the blue berries.
Mirror Plant: Male	*Coprosma* spp.	You'll never know what sex you got when you buy.
Myrtle	*Myrica* spp.	Widely used as a low hedge.
Oaks	*Quercus* spp.	Evergreens are a bigger problem than deciduous.
Olive: Male	*Olea* spp.	Mostly sold as male. Swan Hill is both flower and fruit sterile.
Osage Orange: Male	*Maclura pomifera*	Widely available.
Palms: Male	Many Varieties	Widely planted.
Plane Tree, Sycamore: Male	*Platanus* spp.	Common tree in urban areas.
Sedge	*Carex* spp.	Frequently used as a lawn alternative.
Walnut	*Juglans* spp.	Some species are worse than others.
Willow: Male	*Salix* spp.	Most of this large group of shrubs and trees are profuse producers.

Table 6.3   Animal Pollinated Allergenic Plants

Common Name	Botanic Name	Notes
Almond	*Prunus communis*	One of the few of this genus that is an allergen.
Bearberry, Buckthorn, Coffeeberry, Hollyleaf Redberry	*Rhamnus* spp.	Many varieties, widely used.
Bay: Male only	*Laurus nobilis*	Widely used as a hedge and shade tree.
Bottlebrush	*Callistemon* spp.	Heavy pollen does not travel far.
Cashew	*Anacardium occidentale*	Good crop, but the plant is also poisonous.
Chamomile	*Chamaemelum nobile*	Not everybody is allergic.
Cow Itch	*Lagunaria pattersonii*	Incredible skin irritant, too.
Desert Olive: Male	*Forestiera* spp.	Pollinated by insects.
Fringe Tree: Male	*Chionanthus* spp.	Widely planted for beautiful flowers.
Jatropha	*Jatropha* spp.	Used as a colorful hedge.
Mango	*Mangifera indica*	Edible tropical pollinated by flies and wasps.
Maple (about half of the species)	*Acer* spp.	There are as many allergy-free species as allergy-giving.
Meadow Rue	*Thalictrum* spp.	It is also a good medicinal plant.
Mexican marigold	*Tagetes lemmonii*	This plant drives some people crazy.
Mesquite	*Prosopis* spp.	Native to southwest.
Pepper Tree: Male	*Schinus* spp.	Females produce allergens, too, but not as bad.
Queensland Lacebark	*Brachychiton discolor*	Tiny sticky hairs on leaves and seedpods are a terrible irritant.
Russian Olive, Silverberry (some)	*Elaeagnus* spp.	Common hedge plant. Widely used.
Senecio—many different names (some varieties)	*Senecio* spp.	There are many varieties that cause severe reactions.
Sumac: Male	*Rhus* spp.	Many varieties, widely used.
Tree of Heaven	*Ailanthus altissima*	A weed in much of the U.S.

For a more complete list of plants, if not complete treatment of the subject, refer to Thomas Leo Ogren's book, *Allergy-Free Gardening: The Revolutionary Guide to Healthy Landscaping*. Ten Speed Press. 2000.

MAINTENANCE STRATEGIES

Good maintenance practices will reduce allergens, as it does all types of air pollution. Every strategy mentioned in the section above, Air Pollution, applies here. The strategies below are aimed at reducing the amount of allergenic pollen.

Blowers

Blowers exacerbate allergies and asthma. And they are as much a threat to plants as they are to us. Parking lots can be vacuumed, walkways swept, and sometimes it's okay to leave a little debris to be washed by the rains. Evidence against blowers is so compelling that over 100 U.S. cities and towns have banned or restricted them.

Mowing

Keeping weedy, grassy landscapes mowed will significantly reduce a large source of pollen—most of these plants are wind pollinated. Wild grassy areas require mowing twice a year to remove flower stalks.

Pruning

Simply pruning away to the pollen producing parts of a plant is a surefire way of reducing allergens. Many plants only bloom once a year making this task easy. Furthermore, a typical plant response to pruning is vegetative, rather than reproductive, meaning the plant will produce more pollutant removing leaves and less pollen-producing flowers.

REACTIONS TO INSECTS: ALLERGIC AND OTHERWISE

About 3% of the population has a severe allergic reaction to bee and wasp stings. Up to 1% of those allergic people have a life-threatening reaction. Bee stings cause more deaths than snakebites. Urban landscape design needs to consider this in its planning.

Twenty-one percent of the population has a strong fear of spiders and insects, particularly those that fly and sting. The degree of fear is gender biased; only 15% of males fear insects, whereas 38% of females feel a strong aversion.[14] For many people the sight of bees, beetvles and wasps will induce anxiety, and possibly a panic attack.

What the statement above means is that pollinator gardens should be kept away from areas with a lot of people and vulnerable populations. More specifically, pollinator gardens should be discouraged around hospitals, elderly care facilities, government buildings, libraries, large commercial/retail complexes, and dense downtown areas.

Figure 6.9 **Flying insects, and bees in particular, can cause distress. Many people react to the sight of flying insects with anxiety and panic. Bee careful, for you and others, when making landscape choices.**

Reducing Pesticides

Every year over a 1 billion pounds of pesticides are used in the U.S. Of all the impacts this has on human health the most worrisome is its effect on the health of children. Prolonged exposure has been linked to birth defects, cancer, and various childhood diseases and disorders. A positive link has also been established between pesticide exposure in expecting mothers and a child's chances of autism.[15]

There are four broad categories of pesticides: Fungicides, herbicides, insecticides, and rodenticides. The discussion below only includes herbicides and insecticides, the two most used in urban areas.[16]

Reducing the Use of Herbicides

No landscape is weed-free, but some are freer than others. More likely than not, landscapes with fewer weeds have some of the design and/or maintenance characteristics listed below.

DESIGN AND MAINTENANCE STRATEGIES

Below are the landscape strategies that reduce the number of weeds.

Aggressive Plants

The best defense is often a strong offense. Using plants that can outcompete weeds is one of the greatest time-savers in maintaining a landscape. Below are the plants commonly planted and/or seeded for weed protection.

Planted: Some of the best planted, non-turf ground covers for outcompeting weeds include clover, Convolvulus, dayflower, *Epilobium*, *Geranium*, ivy, morning glory, nasturtium, *Oenothera*, periwinkle, prostrate natal plum, prostrate rosemary, Santa Barbara daisy, St. John's-wort, wild strawberry, yarrow, and non-mowed perennial grasses such a deergrass, fescue and *Stipa*.

Seeded: Some of the plants that can be seeded, and then reseed prolifically on their own, are alyssum, baby blue eyes, blue-eyed grass, California poppy, *Clarkia*, clover, dandelion, flax, forget-me-nots, golden yarrow, goldfields, *Silene*, tidy tips, and yarrow.

*Note: Some of the plants listed above are highly aggressive and a reduction in weeding will be supplanted with an increase in mowing/pruning. Some might even travel to neighboring landscapes.

Creating Barriers

Weed seeds are designed to travel. They flit and tumble-down streets and sidewalks; they hitch rides on birds and lizards. Whether vegetative or constructed, barriers help block the migration of weed seeds.

Constructed barriers might be a small wall or tall fence. It is anything that disrupts the flow of wind. An effective vegetative barrier will be dense; have leaves that are small, sticky or oily; and foliage that is more stiff than flexible. Some of the best barrier plants include arborvitae, *Ceanothus*, coyote brush, cypress, juniper, *Justicia*, lantana, lavender, myrtle, rockrose, rosemary, *Santolina*, sumac, *Teucrim*, and *Westringia*.

Mow or Scrap

It is okay to let weeds grow; in fact, it is often beneficial. Weeds help break up dense soil, enrich poor soil, and attract pollinators. It is rarely beneficial, however, to let weeds go to seed. Letting weeds produce seed guarantees next year's crop will be just as large, if not larger.

If the weeds are high, such as grasses, mow them before they go to seed. If the weeds are low, such as bindweed and spurge, scrap them off the soil. The timing of this task hinges on observation. The goal is to cut back after the plant has flowered but before it has set seed.

Mulches

Controlling weeds with mulch is a universally recommended method of control. Mulch suppresses growth by blanketing existing weeds and preventing incoming seeds from touching soil and rooting. Mulches can be divided between organic or inorganic.

Organic Mulches

Not all mulches are equally effective at suppressing weeds. Finely decomposed mulches are great for growing plants, but not for weed suppression. Recently chipped plant material is great for weed protection, but not for nourishing plants (at least not initially).

Recently chipped material from plants high in oils offers the best suppression. Wood chips of this sort can chemically inhibit germination, bind soils, and help to slow fast-moving water. The most effective mulches for weed suppression are recently chipped *Acacia*, camphor, *Eucalyptus*, juniper, oak, pine, and *Pittosporum*. Large, thick mulches are preferred over fine, thin mulches.

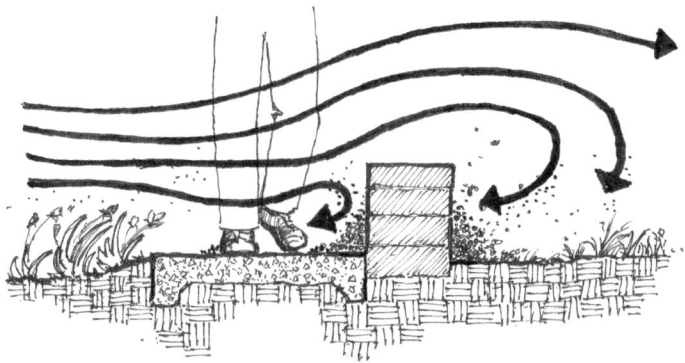

Figure 6.10   A small barrier can help reduce the amount of weed seeds being blown across a property. As wind rushes over a small barrier a wind eddy is created, causing the wind to circle in front and behind the barrier, allowing the debris and weed seeds to settle out.

Inorganic Mulches

Inorganic mulches, such as decomposed granite, gravel and river rock, can help suppress weeds, but do not get rid of them. The biggest benefit of using inorganic mulch is the ease of eradicating weeds. Whatever weeding method you use—be it flaming, pulling, scraping or spraying, inorganic mulches can make the task easier.

Reducing the Use of Insecticides

No landscape is pest-free, but some are freer than others. More likely than not, landscapes with fewer pests have some of the design and/or maintenance characteristics listed below:

- **Plant Compatibility**: One of the surest ways to reduce pests is to make sure that the plants chosen for the landscape are compatible with the environment and irrigation goals. There are many regional guides and references on picking the right plant for the right spot.
- **Good Horticultural Practices**: Irrigating correctly, pruning judiciously and at the right time, and fertilizing only when necessary will ensure plant health, which in turn reduces pest problems.
- **Cleanliness**: A common denominator for pest reduction is cleanliness. That means eliminating the places where pests live and breed. Old wood and leaf piles are removed; overgrown weedy areas are mowed and raked; and areas of storage are pulled out, swept and restacked.
- **Horticultural Diversity**: Having many different types of plants in a landscape ensures that if one plant gets infested, the entire landscape is not at risk. Horticultural diversity also reduces the likelihood of nutrient depletion and damage from climatic extremes. Refer to the section on companion plants below.
- **Successional Diversity**: Succession is the process of one type of plant community transitioning to another type. Higher succession plants that have longer lives benefit from having shorter-lived, lower succession plants around them. See the explanation below.
- **Know When to Quit**: Change is the only constant in living systems, and a landscape is a living system. It is always growing and evolving. What once was full sun may now be shade and soils originally alkaline may become acidic. Plants not compatible to evolving conditions will show signs of stress and pest infestation. At this point, just quit; remove the plant and replace it with one that is more compatible with the evolving conditions.

SUCCESSIONAL DIVERSITY

Successional trajectory is a theoretical model used to explain how landscapes recover from a disturbance, such as fire or flood. As an example: If a wildfire devours a forest, the forest will eventually return, but it would do so by a series of successions. The first plants to sprout after the fire would be the opportunists and pioneers, the low-succession annuals

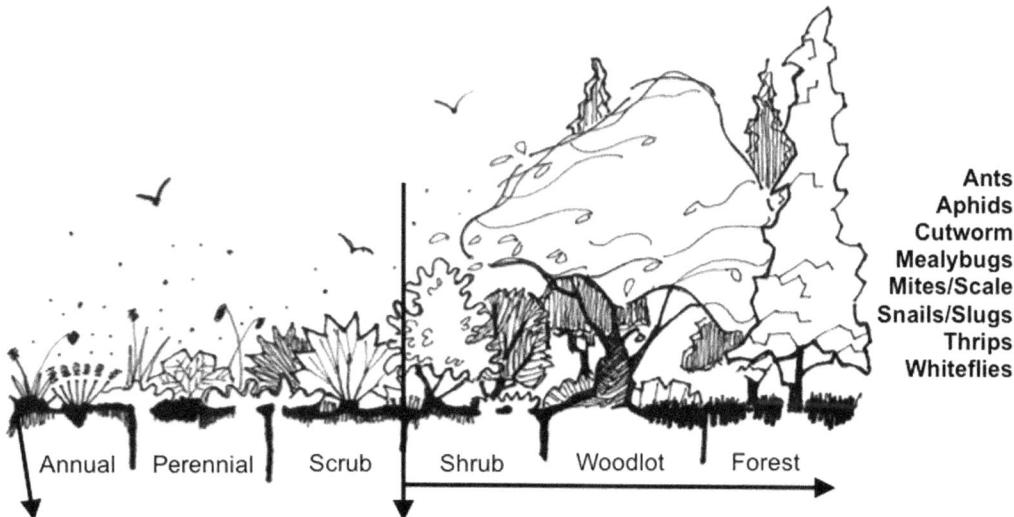

Plants: Angelica, *Artemisia*, berries, borage, carrots, chives, garlic, marigolds, mints, mustards, nasturtium, onions, peppers, petunias, rue, tansy, tobacco, yarrow

Figure 6.11   **Using lower succession plants along with higher succession ones will help the landscape defend itself against pests. Lower succession plants attract beneficial insects, repel unwanted insects, and/or trap unwanted insects.**

and biennials. The next succession would be the perennials, followed by scrub, shrub, mixed evergreen, and finally returning to forest.

Most noxious pest problems occur on the longer-lived, higher-succession plants (food crops are an exception). However, the plants recommended to combat the pests, either by repelling them or attracting their predators, are lower-succession and shorter-lived. Simply put: Higher-succession plants are healthier when accompanied by lower-succession plants.

COMPANION PLANTS

A companion plant is one that improves the health of the plants growing around it. Companion plants can be used to attract pollinators, control pests, enrich soil, or improve the flavor of certain crops. For pest control there are four general categories: Plants that attract a pest's predator, plants that repel pests, plants that trap pests, and plants that are used for concoctions, such as insecticides.

A more detailed discussion is provided in the Landscape Materials chapter, Pest Control section.

Managing Nighttime Light[17]

Night lighting impacts the health of humans. Those living in urban areas with 500,000 people or more are exposed to nighttime lights that are three to six times more intense than those living in small towns or rural areas. Night lighting reduces our ability to produce melatonin, which in turn leads to inadequate sleep, which is linked to obesity, metabolic disturbance, and poorer overall health.

But night lighting impacts far more than humans. It is also highly disruptive to wildlife and disorientates many birds, bats, moths, insects, and nocturnal hunters, such as coyotes. Follow these tips to ensure safety for all:

- Remove excess and redundant lighting—make sure lights have a clear purpose.
- Direct light only to where it is needed, called targeted lighting.
- Use shields on lights to improve targeting.
- Install motion detectors for both in and outdoor lights.
- Replace bulbs with lower illumination.
- Replace bulbs with warmer colors, which have a longer wavelength and are less disruptive than the hyper blue wavelengths of LEDs.
- Use light colored surfaces around lights. The reflection means less illumination is needed.
- Put lights close to the ground along pathways.
- Install fences, build walls or plant hedges around lights to keep it from escaping.

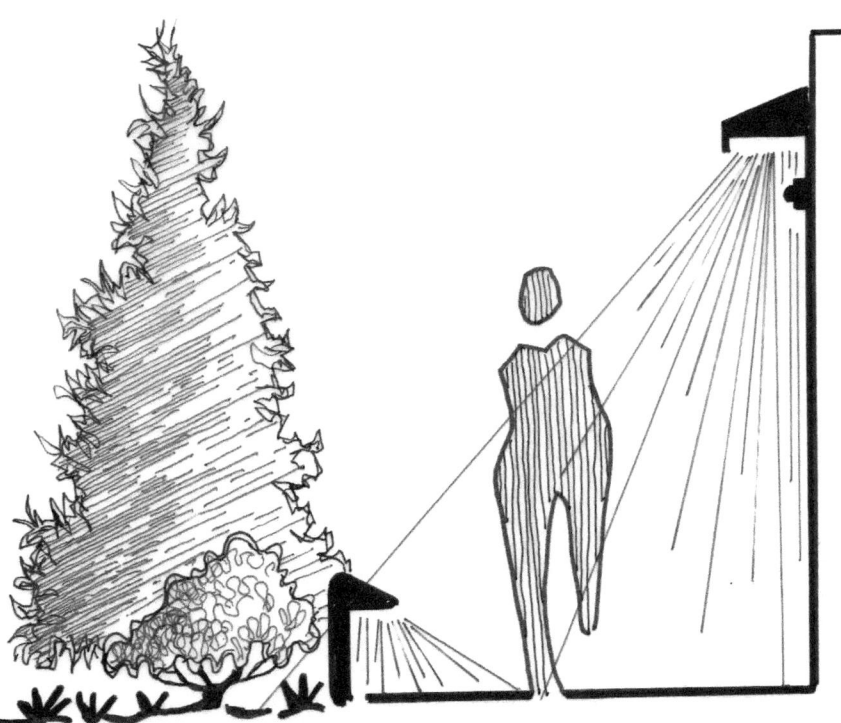

Figure 6.12   **This illustration shows some of the strategies for reducing the impacts of night lighting. Illustrated are down lighting, low-watt bulbs, motion activated lighting, light reflecting surfaces, light close to surfaces, and hedges or walls to block fugitive light.**

Designing Emotional Wellbeing[18]

Humans have a vegetative sweet spot. Certain types of landscapes will lower our heart rate and blood pressure, increase our feelings of security and wellbeing, all of which improves our immunity, morale and longevity.

Landscaping for emotional wellbeing involves the principles of restorative practices. These principles are employed around schools, hospitals, and homes. Below are ten attributes of a landscape that will improve personal and public health.

Notably, none of the ten attributes below come with a plants list. As elegantly stated in *Restorative Commons* (Lindsay Campbell and Anne Wiesen), plants are personal, and a restorative experience will reflect the nature of both the people and place.

Security

People need to feel secure in a landscape. There are two distinct parts to security. First, the visitor must feel buffered from the outside world; they must be shielded from urban chaos and pollutants. Second, they need horizontal visibility. Humans are animals and are constantly scanning their environment for danger. Horizontal visibility and instant recognition of the environment leads to a calming response.

Mobility/Accessibility

A landscape must not only be safe to physically navigate, but must also look like it is. People will feel anxiety about landscapes that do not look safe to traverse. Characteristics

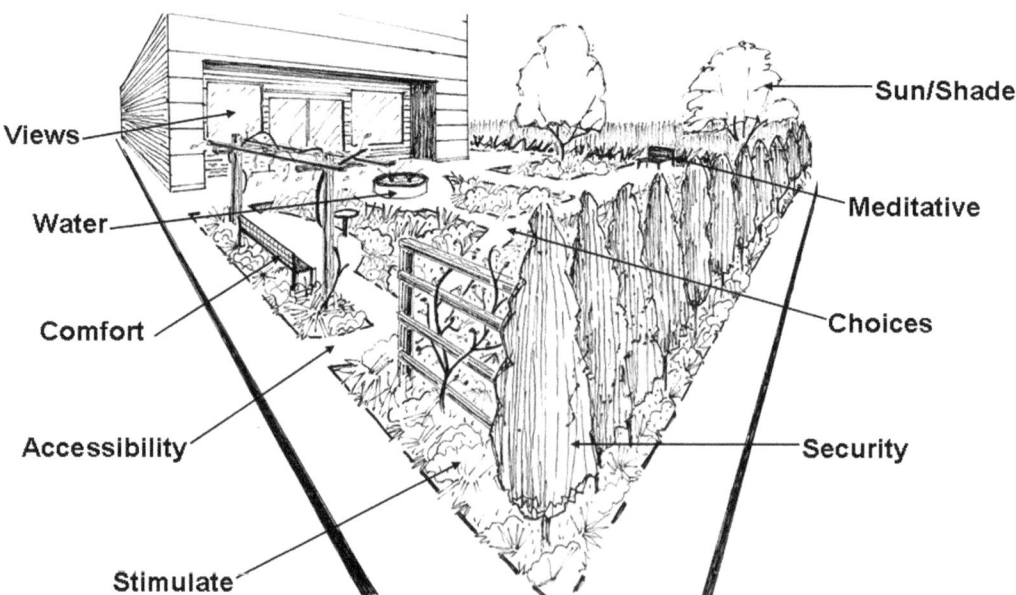

Figure 6.13   A restorative landscape—one that calms and soothes, nourishes and protects—is built around the principles illustrated above.

that produced the greatest feeling of mobility are paths three feet or wider, stable walking surfaces, and curvy and meandering walkways. Also, and importantly, earth-toned walkways are best at calming; stark white creates a glare that causes anxiety.

Choices/Options

People feel good when they have choices, and a good garden will provide many. The goal is to create opportunities for engagement. From the obvious, such as open areas and pathways, to the nuanced, like hidden benches and fruit hanging in trees, each little thing demands a choice: To engage or not. Even lightweight and moveable chairs provide more choices than heavy metal chairs.

Comfortable Sitting

Simply looking at a wooden bench under an old tree increases feelings of wellbeing. Comfort can be seen and experienced. Wood is an ideal material for benches, chairs and seating walls, as its nature is understood and it is never too cold or hot to sit on. Consult a landscape architect's handbook for the proper design of a seating element; only the proper height, seating surface size, and inclination of backrest will calm a person. And importantly, make seating available every 500 feet.

Stimulate Senses

A restorative landscape reaches for its occupants; it interacts with all their senses. Creating a stimulating landscape involves two things: Design and plant selection. In summary, the hardscape must make the benefits of plants accessible to the occupants. Landscape plants must have touchable and engaging foliage, inviting fragrance, showy flowers, and/or enticing fruit.

Sun/Shade

Sounds simple, but not all landscapes allow people to move between cool and warm areas with ease. The design goal is to enable a variety of temperature experiences. A pergola dripping with grapes is the reward for traversing a warm path. Sheltered reflecting walls that absorb the sun's heat provide comfort in a cool wooded environment. Creating areas of sun and shade gives the visitor a choice, which improves the experience.

Water Features

Water features can have a profound impact on feelings of wellbeing because they reach out to four of our five senses. Water is visual and dynamic. It has a distinct smell. It feels familiar and safe. And its sound is predictable yet a bit erratic. Water must be within reach of a person to activate all four senses.

Noise

Science has shown that a predictable noise induces calm, even if it is a little random and chaotic. The most rhythmic and soothing noises in a landscape are natural, such as birds chirping, water falling, and wind rustling leaves.

Meditative Spot

Even if rarely visited, a visually accessible contemplative place in a landscape inspires a calmer state of mind. Whether a lone wooden chair under a weeping tree or a small widening in a path beside a water feature, the goal is to give the heart and mind a place to physically and visually rest.

Views

With a good design people can gain all the benefits of a restorative landscape from inside a structure. They can see the paths and points of rest, the birds and wind in trees. Architecture

Figure 6.14   Shaded, comfortable, and loaded with color and cheer, this landscape was designed to improve the emotional wellbeing of the homeowners.

aimed at emotional health will open a building up to greenery and its dynamic and healing nature.

Resources

BOOKS

Restorative Commons: Creating Health and Well-being through Urban Landscapes. Edited by Lindsay Campbell and Anne Wiesen. USDA Forest Service, Northern Research Station. January 2009. General Technical Report NRS P-39. https://www.fs.usda.gov/research/treesearch/18810

Marcus, Clare Cooper and Sachs, Naomi A. *Therapeutic Landscapes: An Evidence-Based Approach to Designing Healing Gardens and Restorative Outdoor Spaces.* Wiley & Sons, Inc, 2013.

Thomas Leo Ogren, *Allergy-Free Gardening: The Revolutionary Guide to Healthy Landscaping.* Ten Speed Press. June 1, 2000.

WEBSITES

American Horticultural Therapy Association: www.ahta.org
California Horticultural Therapy Network: www.californiahorticulturetherapy.com
International Dark Sky Association. https://www.darksky.org/

Notes

1 Berman, Marc G., et al. "Interacting with Nature Improves Cognition and Affect for Individuals with Depression." *Journal of Affective Disorders*, vol. 140, no. 3, March 2012, pp. 300–305. https://www.ncbi.nlm.nih.gov/pmc/articles/PMC3393816/
    Glausiusz, Josie. "'Mind & Brain/Depression and Happiness—Raw Data' Is Dirt the New Prozac?" *Discover Magazine*, June 13, 2007. https://www.discovermagazine.com/mind/is-dirt-the-new-prozac
    and
    Lowry, C. A., et al. "Identification of an Immune-Responsive Mesolimbocortical Serotonergic System: Potential Role in Regulation of Emotional Behavior." *Neuroscience*, vol. 146, May 11, 2007, pp. 756–772. https://www.ncbi.nlm.nih.gov/pmc/articles/PMC1868963/

2 Thompson, Richard. "Gardening for Health: A Regular Dose of Gardening." *Clinical Medicine*, vol. 18, no. 3, June 2018. https://www.ncbi.nlm.nih.gov/pmc/articles/PMC6334070/

3 Feldmar, Jamie. "Gardening Could be the Hobby that Helps you Live to 100" *BBC*, December 10, 2018. https://www.bbc.com/worklife/article/20181210-gardening-could-be-the-hobby-that-helps-you-live-to-100
    Li, ZH-Hoa, et al. "Leisure Activities and All-Cause Mortality Among the Chinese Oldest-Old Population: A Prospective Community-Based Cohort Study." *Journal of American Medical Director's Association*, vol. 6, June 21, 2020, pp. 713–719. https://pubmed.ncbi.nlm.nih.gov/31588027/
    and
    McCallum, John, et al. "Delaying Dementia and Nursing Home Placement: The Dubbo Study of Elderly Australians over a 14-Year Follow-Up." *Annals of the New York Academy of Sciences*, October 2007, pp. 121–129. https://pubmed.ncbi.nlm.nih.gov/17986578/

4   Correia, Andrew W., et al. "The Effect of Air Pollution Control on Life Expectancy in the United States: An Analysis of 545 US Counties for the Period 2000 to 2007," *Epidemiology*, vol. 1, January 24, 2013, pp. 23–31. https://pubmed.ncbi.nlm.nih.gov/23211349/

5   Gillespie, Terry J and Brown, Robert D. "Modifying Air Quality." *Landscape Architectural Graphic Standards*. Ed. Hopper, Leonard, J. John Wiley & Sons. 2007. p. 99.

6   Xing, Yang, and Brimblecombe, Peter. "Role of Vegetation in Deposition and Dispersion of Air Pollution in Urban Parks." *Atmospheric Environment.*, vol. 201, March 2019, pp. 73–83. https://ui.adsabs.harvard.edu/abs/2019AtmEn.201...73X/abstract

7   Bell, Ryan and Wheeler, Jennie. "Talking Trees: An Urban Forestry Toolkit for Local Governments." *ICLEI: Local Governments for Sustainability.* November 2006. http://www.milliontreesnyc.org/downloads/pdf/talking_trees_urban_forestry_toolkit.pdf

8   Bell and Wheeler. *Talking Trees.* ICLEI. November 2006. See above.
      and
      "SelecTree: A Tree Selection Guide." Urban Forest Ecosystems Institute, NRES Department, California Polytechnic State University, San Luis Obispo, CA. http://selectree.calpoly.edu

9   Virginia Gorsevski, et al. "Air Pollution Prevention through Urban Heat Island Mitigation: An Update on Urban Heat." *Island Pilot Project, Draft Report.* Lawrence Berkeley National Laboratory, Berkeley, January 6, 1998. https://www.osti.gov/biblio/8871
      Nowak, David J. and Heisler, Gordon M. "The Effects of Urban Trees on Air Quality." National Recreation and Park Association, 2010. https://www.nrpa.org/globalassets/research/nowak-heisler-research-paper.pdf

10 Wang, Huixia, et al. "Efficient Removal of Ultrafine Particles from Diesel Exhaust by Selected Tree Species: Implications for Roadside Planting for Improving the Quality of Urban Air." *Environmental Science & Technology,* vol. 53, no. 12, June 18, 2019, pp. 6906–6916. https://ui.adsabs.harvard.edu/abs/2019EnST...53.6906W/abstract

11 Takahashi, Mise and Morikawa, Hiromichi. "Nitrogen Dioxide is a Positive Regulator of Plant Growth." *Plant Signaling and Behavior,* vol. 9 no. 2, February 13, 2014. https://pubmed.ncbi.nlm.nih.gov/24525764/
      Jiechen, Wang, Yue, Wang, Huihui, Zhang, Dandan, Guo, and Guangyu, Sun. "Atmospheric Nitrogen Dioxide at Different Concentrations Levels Regulates Growth and Photosynthesis of Tobacco Plants." *Journal of Plant Interactions* vol. 16 no. 1, 2021, pp. 422–431. https://pubag.nal.usda.gov/catalog/7609912

12 Cook-Anderson, Gretchen. "Urban Heat Islands Make Cities Greener." *NASA.* June 6, 2004. https://www.nasa.gov/centers/goddard/news/topstory/2004/0801uhigreen.html

13 "SelecTree: A Tree Selection Guide." Urban Forest Ecosystems Institute, California Polytechnic State University, San Luis Obispo, CA. http://selectree.calpoly.edu/
      and
      Ogren, Thomas Leo, *Allergy-Free Gardening: The Revolutionary Guide to Healthy Landscaping.* Ten Speed Press. June 1, 2000.

14 Brewer, Geoffrey. *Snakes Top List of Americans' Fears: Public Speaking, Heights and Being Closed in Small Spaces also Create Fear in Many Americans.* March 19, 2001.Gallup News Service, August 12, 2001. https://news.gallup.com/poll/1891/snakes-top-list-americans-fears.aspx
      Orth, Taylor. "Three in 10 Americans Fear Snakes — And Most Who Do Fear Them a Great Deal." *YouGovAmerica,* June 17, 2022. https://today.yougov.com/topics/society/articles-reports/2022/06/16/americans-fear-snakes-heights-spiders-poll

15 Schafer, Kristin S. and Marquez, Emily C. "A Generation in Jeopardy: How Pesticides are Undermining our Children's Health & Intelligence." Pesticide Action Network. October 2012. https://www.panna.org/sites/default/files/KidsHealthReportOct2012.pdf

16 **Fantastic Resources for Pest Control**
      Kent, Douglas. *California Friendly: A Maintenance Guide for Landscapers, Gardeners and Land Managers.* Muni, 2017. https://www.bewaterwise.com/assets/ca-friendly-maintenance-book.pdf
      California Master Gardener Handbook, 2nd ed. Ed. Pittenger, Dennis R., University of California, Agriculture and Natural Resources. September 2017.

Pests of Landscape Trees and Shrubs: An Integrated Pest Management Guide. 3rd ed. Ed. Dreistadt, Steve H. and Flint, Mary Louise. University of California: Agriculture and Natural Resources. Publication 3359. 2016.

Invasive Plants of California's Wildlands. Ed. Bossard, Carla C., et al. University of California Press. 2000. https://www.cal-ipc.org/resources/library/publications/ipcw/

Holloran, Pete, et al. *The Weed Workers' Handbook: A Guide to Removing Bay Area Invasive Plants. The Watershed Project and California Invasive Plant Council.* 2004. https://www.cal-ipc.org/product/the-weed-workers-handbook/

17   International Dark Sky Association. https://www.darksky.org/

18   **Excellent Resources for Restorative Landscapes**

Restorative Commons: Creating Health and Well-Being through Urban Landscapes. Ed. Campbell, Lindsay and Wiesen, Anne. USDA Forest Service, Northern Research Station. January 2009. General Technical Report NRS P-39. https://www.fs.usda.gov/research/treesearch/18810

Marcus, Clare Cooper and Sachs, Naomi A. *Therapeutic Landscapes: An Evidence-Based Approach to Designing Healing Gardens and Restorative Outdoor Spaces.* Wiley & Sons, Inc, 2013.

Chapter 7

Self-Care

Plants that provide care should be planted outside every entryway. They should be planted around every window, every patio, and every outdoor living area too. The plants listed in this chapter can provide incredible care if they are within reach of the people that need them.

This chapter lists a myriad of plants that can be harvested and brought indoors to use for bouquets, fragrances, medicines, soaps, and shampoos. There is more on the topic of health and wellbeing in the chapters on Craft, Food, Public Health and Thermal Comfort.

Importantly, this chapter does not provide recipes. Its job is to illustrate what can be found or grown in urban landscapes. At the end of this chapter is a list of books that provide the remedies and specifics of crafting health and wellbeing.

Quick Overview of Self-Care

Uses: Whether bouquets or cleansing sticks, soaps or teas, the everyday use of self-care plants and practices can increase feelings of wellbeing, improve confidence, and reduce the amount of time spent with doctors.

Costs: Low to medium.

Difficulty: Self-care can be as simple as growing herbs and making tea.

Best Scale: The best place for landscapes that care for people is around homes, care facilities, nature centers, schools, and small business. These are the properties where people congregate, spend a significant amount of time, and have access to the tools of preparation, such as a cutting board, sink, and burner.

DOI: 10.4324/9781003369752-7

ERoEI: If any of these historic remedies work for you, the return on investment is great. If any of the medicine comes from a spontaneous plant, like cheeseweed or dandelion, the ERoEIs are phenomenal.

Pros: We evolved from the Earth, are nourished by the Earth, and when we embrace our historic relationship with Earth, we are healthier.

Cons: The largest disadvantage is the time and attention it takes to grow and create some of these remedies. Just like growing food or timber, the art of self-care is in the cultivating, harvesting, and use.

Bouquets

Bringing flowers indoors is far more than beauty. Flowers bring nature, color and fragrances indoors, which is turn reduces stress, boost moods, increase memory, help relationships, and inspire healing.[1]

While there are hundreds of plants that produce cut flowers, the ones below have certain characteristics. Their flowers are long lasting, they are profuse bloomers, they require few resources to grow, and many naturalize.

Figure 7.1   **Growing cut flowers in urban areas is as much a spectacle as it is an economic harvest. These fields generate revenue through tourism and the selling of cut flowers and propagules (bulbs). Picture taken at the Flower Fields, Carlsbad, CA.**

Good Cut Flowers and Foliage for Urban Gardens

Notes:

- **D** denotes the plants that are ideal for dried arrangements. They are slow to lose their petals, color and/or rigidity.
- **F** denotes the plants that have a fragrance.
- **N** denotes the plants that can be found naturally growing in urban areas.

African daisy, *Arctotis* spp.
Agave spp. F, N
Aloe spp. F
Amaranth, *Amaranthus*. D, N
Amaryllis spp. N
Artichoke. D
Aster spp. N
Banksia spp. D
Basil, *Ocimum basilicum*. F
Blanket flower, *Gaillardia* spp. D
Broom, *Cytisus*, *Genista* and *Spartium* spp. D, N
Buckwheat, *Eriogonum* spp. F, N
Burdock, *Arctium* spp. D, N
Butterfly bush, *Buddleja* spp. F
Cattail, *Typha* spp. N
Coastal sunflower, *Encelia californica*. N
Checkerblooms, *Sidalcea* spp. N
Cherry, *Prunus* spp. N
Columbine, *Aquilegia* spp. N
Cone bush, *Leucadendron* spp. D
Coneflower, *Echinacea* spp. N
Coral bells, *Heuchera* spp. N
Tickseed, *Coreopsis* spp. N
Cosmos spp.
Cotoneaster spp.
Currant and gooseberry, *Ribes* spp. N
Daffodils, *Narcissus* spp. F, N
Daylily, *Hemerocallis* hybrids
Eucalyptus spp. D, F, N
Feverfew, *Chrysanthemum parthenium*
Foxglove, *Digitalis purpurea*. N
Freesia spp.
Gaura lindheimeri
Grevillea spp.
Heliotrope, *Heliotropium* spp. D
Horehound (white), *Marrubium vulgare*. D, F, N
Iris spp. N

Kangaroo paws, *Anigozanthos* spp.
Larkspur, *Delphinium* spp. D, N
Lavender, *Lavandula* spp. D
Lilac, California, *Ceanothus* spp. F, N
Lilac, *Syringa* spp. F, N
Magnolia, *Magnolia* spp. F
Matilija poppy, *Romneya* spp. N
Mugwort, *Artemisia ludoviciana ssp. albula*. F, N
Mullein, common, *Verbascum thapsus*. N
Onion, *Allium* spp. F, N
Passionflower, *Passiflora edulis* N
Pearly everlasting, *Anaphalis margaritacea*. D, N
Peppergrass, *Lepidium* spp. (taller varieties). D, N
Pride of Madeira, *Echium* spp. N
Protea spp. D
Redbud, *Cercis* spp. N
Red hot poker, *Kniphotia uvaria*
Rudbeckia spp. D, N
Rose, *Rosa* spp. F, N
Rosemary, *Salvia rosmarinus*. D, F
Sagebrush, *Artemisia* spp. D, F, N
Sea holly, *Eryngium planum*. D, N
Sea lavender, *Limonium vulgare*. D, N
Sunflower, *Helianthus* spp. N
Toyon, *Heteromeles arbutifolia*. N
Thistle, *Cirsium* spp. and *Silybum* spp. D, N
Vetch, *Vicia* spp. N
Throatwort, *Trachelium caeruleum*. D, F
Woad, *Isatis tinctoria*. N
Woolly blue curls, *Trichostema lanatum*. F, N
Yarrow, *Achillea* spp. D, F, N
Yucca spp. and *Hesperoyucca whipple*. F, N
Zinnia spp. d

Fragrances

Releasing fragrances is a fantastic way to boost morale and mood. Fragrances can embolden, like bay; can sooth, like jasmine; or can arouse, like sweet violet. Highlighted below are the terms used in fragrance making and the best plants to either grow yourself or find spontaneously growing.

Most homemade fragrances are short lived and should be used within weeks of making them.

Figure 7.2   **A little does a lot. A few sprigs of spearmint and one fragrant rose is just enough to invigorate this workstation, ultimately boosting moods and morale. If close at hand, keep bouquets small and fresh, replacing them every 3 to 5 days.**

Nomenclature

Base Notes: The least volatile compound and the longest lasting note of a fragrance. It provides the foundation and harmony.

Carrier Oil: Carrier oils have two jobs. First, they are the medium that allows the aromatics to stick to your skin. Second, they help reduce irritation that may occur with some fragrances. Carrier oils are mostly unscented, but not always. The common carrier oils include avocado seed, castor, fragmented coconut, grape seed, jojoba wax, olive, rapeseed, rose hip and sweet almond.

Cleansing Stick: Fragrant dried material bound together and burnt. Also called smoke cleansing. The term smudge stick is similar but implies a certain set of rituals, which include acts of reciprocation.

Cologne: A fragrance with a low concentration of aromatics and essential oils. Its scent lasts no longer than 2 hours.

Essential Oils: A liquefied and concentrated version of something. Common methods of extracting plant oils include solvents (alcohols, glycerin, oils, vinegars), maceration, steam distillation, and water distillation.

Fixative: A solution that helps all the oils blend. One hundred proof vodka and 190 proof alcohols are common. Beeswax is also used for thicker concoctions.

Fragrance: A smell that has been captured and its release can be controlled.

Glycerin: A term for glycerol, a colorless, odorless liquid that is non-toxic and has antimicrobial and antiviral properties. Glycerin is used as a humectant (to reduce moisture loss) and solvent. Glycerin can be derived from both animal and plant oils.

Middle Notes: The most dominate scent of a fragrance.

Perfume: The most intense fragrance. It has a high percentage of essential oils, it is generally thick, and its scent is the longest lasting.

Preservative: A solution that helps extend the life of a fragrance. Vitamin E and grapefruit seed extract are common.

Smudge Stick: A term used by Northern American indigenous people to describe the rituals of harvesting, preparing, and burning native plants.

Top Notes: The first and most fleeing scent. Top notes evaporate quickly.

THE SYMPHONY OF FRAGRANCE

An enticing and lingering fragrance is comprised of three notes: Top, middle and base. The top is the first smell and most fleeting; the middle is the core and most penetrating; and the base accentuates and is the longest lasting. Although percentages vary widely, a good first-time recipe is 30% top notes, 50% middle, and 20% base.

Methods of Utilization

There are a variety of ways to harness the fragrance in plants. Below are the most common.

- **Extract**: The common extracts are distillation and infusion. Distillation is the process of vaporizing reactive compounds, which are then captured, condensed and separated. Distillation occurs through boiling, CO_2 extraction, chemicals, and steaming. Infusion occurs where the material is saturated in a solvent, such as alcohol or oil, and the beneficial compounds are leached. Extracts are used for concoctions, such as candles, disinfectants, medicines, perfumes, and soaps.

- **Burn**: Burning plants to fill a space with fragrance is instinctive human behavior. It is great for odor removal and reinvigoration. Cleansing sticks, which are fragrant dried material tightly bound together, are easy to make. Some of the common plants for burning include bay, camphor, cedar, cypress, *Eucalyptus*, juniper, lavender, mugwort, mullein, pine, rosemary, Russian sage (*Perovskia atriplicifolia*), and a wide variety of the fragrant *Salvias*, most notably the white sage (*Salvia apiana*). Cleansing sticks are often, and mistakenly, called smudge sticks.
- **Natural and Dried**: Scented and dried material can be used to great effect in bouquets, potpourri and hanging bundles in damp closed areas, such as bathrooms. Some of the best plants include bay, cedar, *Eucalyptus*, juniper, lavender, and scented geraniums.
- **Natural and Fresh**: Many plants have strong fragrance when they are first cut but lose their scent quickly. Generally, flowers are short-lived fragrances. Some of the most fragrant flowers are heliotrope, jasmine, lilac, mock orange, rose, and tuberose. Leaves, on the other hand, are longer lived, and include plants like bay, juniper, pine, rosemary, and white sage.
- **Simmer**: Lightly simmering scented material in water is a good way to fill a space with fragrance. Plants ideal for simmering include *Eucalyptus*, camphor, and pine.

Figure 7.3   **Cleansing sticks are a great way to infuse an area with a dense and invigorating fragrance. Pictured are common urban sticks and include incense cedar, blue gum *Eucalyptus*, tam juniper, French lavender, California mugwort, Canary Island pine, rosemary, and tea tree.**

Common Fragrance Plants

Legend
C: Can be cultivated
F: Can be naturally found

Basil, *Ocimum basilicum*: Use leaves. Provides top and middle notes. Many types of scents, including chocolate, cinnamon and lemon. Easy to grow. C

Bay: Use leaves. There are two varieties: California bay/myrtlewood (*Umbellularia californica*) and sweet bay (*Laurus nobilis*). Use myrtlewood cautiously, as not everyone responds positively. Provides middle and base notes. Abundant. C, F

Bergamot, wild, *Monarda fistulosa*: Use leaves and flowers. Provides top and middle notes. C, F

Blue gum Eucalyptus, *Eucalyptus globulus*: Use leaves. Provides top and middle notes. Abundant. C, F

Camphor, *Cinnamomum camphora*: Use leaves and distillation from wood. Base notes. Abundant. C, F

Cedar, incense, *Calocedrus decurrens*: Use leaves, resin and sap. Provides base and middle notes. Can be abundant in the western facing foothills along the west coast. C, F

Cedar, western red, *Thuja plicata*: Use leaves and resin. Provides base and middle notes. C, F

Citrus: Use buds, leaves, petals, and rinds. Provides top notes. The most used are lemon, lime and orange. Abundant in frost-free areas. C

Cypress, Italian, *Cupressus sempervirens*: Use leaves. Provides middle and base notes. Abundant. C

Fennel, *Foeniculum vulgare*: Use leaves and seeds. Provides top and middle notes. Abundant. C, F

Fir, *Abies* spp.: Use needles. Provides middle and base notes. Every fir produces a unique scent. Douglas Fir, *Pseudotsuga menziesii*, is frequently used. C, F

Gardenia, *Gardenia* spp.: Use flowers. Provides top and middle notes. Use only the most fragrant varieties.

Geranium, scented, *Pelargonium* spp.: Use leaves. Provides top and middle notes. Scents include apple, chocolate, citronella, lemon and rose. Easy to grow. C

Heliotrope, *Heliotropium* spp.: Use flowers. Provides top notes. Can be abundant. C

Jasmine, *Jasminum officinale*: Use flowers. Provides high and middle notes. Not common but can be grown in mild winter areas. C

Juniper, *Juniperus* spp.: Use berries, leaves, and distillation from wood. Provides middle and base notes. Abundant. C, F

Lavender, *Lavandula* spp.: Use leaves or flowers. Provides top notes. Use only the most fragrant varieties. Abundant. C

Lemon, *Citrus limon*. Use petals, fruits, and rinds. Provide top notes. C

Lemon balm, *Melissa officinalis*: Use leaves. Provides top notes. Easy to grow. C

Lemon grass, *Cymbopogon citratus*: Use leaves. Provides middle notes. Can be aggressively abundant. C

Lemon scented gum, *Corymbia citriodora*: Use leaves. Provides top and middle notes. Abundant. C, F

Lemon verbena, *Aloysia citriodora*: Use leaves. Provides top notes. Easy to grow. C

Lilac, *Syringa vulgaris*: Use flowers. Provides top and middle notes. Common in areas with a pronounced chill. C, F

Magnolia, Southern, *Magnolia grandiflora*: Use petals. Provides middle notes. C, F

Mint, *Mentha* spp.: Use leaf. Provides top notes. The most used include pennyroyal, peppermint, and spearmint. Abundant. C, F

Mock orange, *Philadelphus* spp. Use flowers. Provides top and middle notes. Can be abundant. C, F

Peppermint tree, *Agonis flexuosa*: Use leaves. Provides middle notes. C

Pine, *Pinus* spp.: Use needle and distillation from wood. Provides base notes. Abundant. C, F

Rose, *Rosa* spp.: Use petals. Provides middle notes. Use only the most fragrant varieties. Abundant. C, F

Rosemary, *Salvia rosmarinus*: Use leaves. Provides middle and base notes. Abundant. C

Sage, *Salvia* spp.: Use flowers or leaves. Provides top to base notes. Any of the aromatic sages will work, such as black, Cleveland, cooking, purple and white sages. Abundant in the southwest. C, F

Sagebrush, *Artemisia* spp.: Use leaves and stems. Provides middle notes. Any of the aromatic sages will work, such as big, California, mugwort and wormwood. Abundant in western U.S. C, F

Tea tree, *Melaleuca alternifolia*: Use leaves. Provides middle notes. Tolerates a wide range of conditions and can be easy to grow. C

Thyme, *Thymus* spp.: Use leaves. Provides top and middle notes. Common in Mediterranean gardens and grocery stores everywhere. C

Tuberose, *Polianthes tuberosa*: Use flowers. Provides powerful middle notes. Grown throughout the U.S, but mostly in the southern states. C

Violet, sweet, *Viola odorata*: Use flowers. Provides high notes. Common garden plant. C, F

Health Remedies

Almost everything described in this book will improve personal and public health. This section is more targeted. It describes which plants alleviate specific ailments, and it describes the kind of remedies that can be made from these plants.

Importantly, this section only highlights plants and their concoctions. It does not provide recipes. This is one discipline where details greatly matter, and sometimes gravely. Not everyone reacts the same—*so please, always consult multiple resources, talk to your physician, and engage with others when learning the use of natural remedies.*

Common Non-Industrial Remedies[2]

Listed below are the types of treatments that can be concocted with a bountiful garden, a serviceable kitchen, and ordinary cooking skills. The treatments include extracts, topical treatments, drinks, and aromatherapy.

EXTRACTS

Whether liquid or powder, an extract is the product resulting from the removal or capture of the desirable qualities of a plant. Extracts are created with any combination of boiling, burning, grinding, leaching, macerating, screening, seeping, steaming, and using solvents. Common extracts include decoctions, essential oils, infusions, and tinctures. Extracts are also called concentrates.

Decoction: Simmering a plant for a long period of time to extract its water and heat soluble qualities. Decoctions are generally brewed from coarse materials, such as bark, root and stem.

Essential Oils: A liquefied and concentrated version of something. Common methods of extracting plant oils and terpenes include carbon, chemicals, maceration, steam, and water distillation. Essential oils are generally captured from finer material, such as flowers, leaves and twigs.

Infusion: Saturating a plant in a liquid to extract its essential qualities. The liquid is the solvent and is generally alcohol, oil, vinegar, or water. Infusions can take minutes to make, like teas, or months, like tinctures.

Tincture: Plant material is saturated in alcohol. Tinctures may or may not include water. Because only the strongest alcohol will do, tinctures are administrated in drops, not tablespoons or cups.

TOPICAL TREATMENTS

There are many types of homemade topical treatments. They are crafted to heal a wide range of maladies, whether they be external or internal.

Balm: Balms are the thickest type of topical application. They have a high concentration of beeswax, along with oil and additives. Balms do not penetrate the skin like other treatments and act like a protective shield, making them ideal for chapped lips and diaper rashes.

Cream: Creams are mostly oil, an emulsifier (like beeswax), and a little water. They are crafted to heal body and joint aches, burns, dehydration, dryness, eczema and scarring. Because of their viscosity, creams provide good skin protection and hydration. Creams are not applied to open injuries and sores.

Liniment: Liniments are designed to be vigorously rubbed into the skin. They are used for joint and muscle aches and often include evaporating agents, such as alcohol, to create a penetrating cooling effect. Bay, camphor, cinnamon, *Eucalyptus* and ginger are common in liniments because they also provide cooling and relief.

Lotion: Lotions are lightly applied to unbroken skin. They have higher water content than creams and are more readily absorbed by the skin. They are used to moisturize, soothe, and soften.

Poultice: Poultices treat bone, joint, muscle and skin ailments. They are crafted to reduce inflammation and pain, heat or cool an area, or help skin recover from abrasion or insect bites. They are a pulpy wet mass made from a variety of healing plants, oils and additives, such as clay, flour, onions and potatoes. A wrap is often used to hold the poultice in place.

Ointment: Ointments are like a salve but with more medicine. Ointments address a specific problem, such as an infection. A doctor would prescribe an ointment whereas a home cook would make a salve.

Salve: Salves are a combination of beeswax, oil and additives. Salves are softer than balm, but thicker than ointments. They, along with balms, do not contain water and are long lasting. Salves can be mixed with medicine, like antiseptics, and are softer than balms and can penetrate the skin. They are used for many skin conditions, such as burns, dryness and rashes.

Scrub: Scrubs are used to cleanse and exfoliate skin. Oils are mixed with coarse materials and fragrances to create scrubs. Some of the exfoliating materials are bentonite clay, oat, salt, sunflower and sugar. Some of the healing agents added to scrubs include extracts, honey and yogurt.

Wrap: Wraps not only hold a poultice in place, but they can also be medicine as well. Whole leaves of medicinal plants, slightly broken and/or saturated, are wrapped around the injured area. Some of the healing leaves that are used for wrapping include curly dock, fennel, hoja santa, mustard, plantain, and yerba mansa. A wrap is also called a compress.

DRINKS

Creating drinks for healing and wellbeing is easy and effective.

Elixir: An elixir is a concoction of water, healing plants and minerals. Elixirs are medicine and used to treat allergies, colds, cramps, fevers and other maladies. They can also be used to boost energy, restore electrolytes, and increase feelings of wellbeing.

Tea: One of the simplest forms of extraction. Making tea involves harvesting, processing, boiling, and seeping. Teas are a powerful medicine and many of the recommendations further below involve teas. Refer to the Food chapter for a list of other good urban tea plants.

Tonic: Tonics are a restorative drink and a concoction of water, decoctions, infusions, and/or other additives. They provide an overall boost and are used to help ease allergies, increase energy, restore electrolytes, provide nutrition, reduce stress, and improve feelings of wellbeing. Tonics generally do not address a specific malady, such as a fever, as an elixir would.

AROMATHERAPY AND SENSUAL REMEDIES

Either applied topically or released to fill the air with scents, aromatherapy relies on releasing fragrance, whether naturally (like cut flowers), burning (like cleansing sticks), or through essential oils (like bath oils, lotions and salves).

Figure 7.4   From intense shade and ease of care to many external and internal health remedies, Tea tree, *Melaleuca alternifolia*, is a fantastic urban tree for personal health and self-care. It can be burnt, distilled into an essential oil, and used in bouquets. Its fragrance is invigorating.

Plant Lists

There are hundreds of plants that can be found or cultivated for personal wellbeing. The plants in the abbreviated list below were selected for the following qualities:

- Provide reliable remedies for everyday ailments.
- Require little energy to grow.
- Are easily propagated.
- Can be used with ease.

HANDY MEDICINAL PLANTS TO HAVE AROUND

The lists below are neither a definitive inventory nor a recommendation to use. *Always consult multiple resources, talk to your physician, and engage with knowledgeable herbalists while learning the use of natural medicines.*

Spontaneous Vegetation

Bidens. *Bidens* spp.: Antiseptic, good for digestive and respiratory health, and calming. Use leaves.

Bitter and prickly lettuce, *Lactuca virosa* and *L. serriola*: Provides pain relief. Use leaves and stalks of mature and flowering plants.

Blackberry, *Rubus* spp. Anti-inflammatory and antioxidant. Use berry.

Cat's Ear, *Hypochaeris* spp. Reduce allergy symptoms and improves general health. Use leaves and roots.

Cheeseweed, *Malva parviflora*: Used internally and externally. Internally, it is an anti-inflammatory and antioxidant, and helps clear lungs and settle stomachs. Externally, and as a wash, it helps heal skin Use leaves and roots.

Chicory *Cichorium intybus* Antioxidant and good for digestion and energy. Use leaves and roots.

Chickweed, *Stellaria media*: Anti-inflammatory, antioxidant and great for digestive health and healing skin. Use leaves, flowers, and stalks.

Curly dock, *Rumex crispus*: Gentle laxative, heals skin, and supports overall health. Use leaves, roots, and seeds.

Dandelion, *Taraxacum* spp.: Anti-inflammatory and antioxidant, improves digestive and reproductive functions, heals skin, and reduces depression and pain. Use leaves, flowers, and roots.

Epazote (Mexican tea), *Dysphania ambrosioides*: Reduces flatulence, improves circulation, and may reduce asthma. Use leaves and seeds.

Fennel, *Foeniculum vulgare*: Anti-inflammatory and antioxidant, improves digestive functions, and helps open airways. Use leaves, flowers, seeds, and bulbs.

Fireweed, *Epilobium angustifolium*: Anti-inflammatory and used to improve digestive function. Use leaves and flowers.

Horehound (white), *Marrubium vulgare*: An expectorant good for colds and coughs, improves appetite. Use leaves.

Lambsquarter and Goosefoot, *Chenopodium album* and *C. murale*: Improves digestive and overall health. Use leaves and seeds.

Licorice, *Glycyrrhiza* spp.: Anti-inflammatory and improves digestive and respiratory health. Use roots.

Mullein, *Verbascum thapsus*: Anti-inflammatory, antiseptic and expectorant, and used for digestive, respiratory and skin health. Use leaves and flowers.

Mustard, black, *Brassica nigra*: Anti-inflammatory and used for bone and digestive health. Use leaves, flowers, and seeds.

Pineapple weed, *Matricaria matricarioides*: Used for calming and as a sedative. Use flower heads.

Plantain, common, *Plantago major*: Used internally for digestive health and externally for healing skin. Use leaves, roots, and seeds.

Prickly pear, *Opuntia* spp.: Helps heal skin and aids in digestive health. Use pads (nopales) and flowers.

Purslane, common, *Portulaca oleracea*: Used to boost immunity, heal skin, and aid in digestion. Use leaves and seeds.

Sea Fig and Hottentot Fig, *Carpobrotus chilensis* and *C. edulis*: Used externally to heal skin. Use leaves.

Selfheal, *Prunella vulgaris*: Internally it reduces symptoms of colds and fevers, externally is helps heal skin. Use leaves.

Sow thistle *Sonchus* spp.: Good for digestive health, some pain relief, and removes warts, Use leaves, sap, stalks, and roots.

Stinging nettle, *Urtica* spp.: Good for allergies, digestion, skin problems, and overall wellbeing. Use leaves and roots.

Violet, sweet, *Viola odorata*: Wonderful for digestive and respiratory health, and some relief from headaches and pain. Use leaves and flowers.

Yarrow, *Achillea millefolium*: Enhances digestive functions, reduces severity of colds and flu, and used externally for skin health. Use leaves and flowers.

Yerba Buena, *Clinopodium douglasii*: Used internally for colds, fevers, and digestive problems, externally for skin problems. Use leaves.

Yerba mansa, *Anemopsis californica*: Used to reduce inflammation and pain. Use leaves and roots.

Yerba santa, *Eriodictyon californicum*: Expectorant with many respiratory benefits, including allergies. Use leaves.

Figure 7.5 **Pineapple weed (*Matricaria matricarioides*) is a common weed throughout the U.S. and has been used for centuries to reduce anxiety and improve sleep.**

Cultivated Plants

Aloe vera, *Aloe vera barbadensis miller*: Internally it aids digestion, externally it heals skin. Use leaves.

Amaranth, *Amaranthus cruentus*: Antioxidant that aids in general wellbeing. Use leaves and seeds.

Beebalm, *Monarda* spp.: Used for digestive health and colds/flus and helps calm. Use leaves.

Borage, *Borago officinalis*: Anti-inflammatory that aids digestion, respiratory health, and restlessness. Used externally for skin health. Use leaves, flowers, and seeds.

Burdock, *Arctium* spp.: Used as a detoxifier and enhances overall health. Use roots, leaves, flowers, and seeds.

Chamomile, German, *Matricaria recutita*: Anti-inflammatory that improves digestion, anxiety and restless, and the symptoms of colds and fevers. Externally it helps heal skin. Use flower heads.

Chastetree, *Vitex agnus-castus*: used to reduce severity of PMS and menopause and may boost fertility. Use leaves and berries. C

Chia, *Salvia columbariae*: Antioxidant that improves overall and digestive health. Use seeds.

Curry leaf, *Murraya koenigii*: Great for overall health and digestion with mild anti-inflammatory properties. Use leaves.

Echinacea, *Echinacea purpurea*: Enhances immunity and respiratory system. Externally it is antibacterial and heals skin. Use leaves, flowers, and roots.

Ginger, *Zingiber officinale*: Enhances digestive and respiratory systems and may calm and relieve pain. Use roots.

Juniper, *Juniperus* spp.: Good for respiratory health, digestive processes, and calming. Use berries and leaves.

Lavender, *Lavandula* spp.: Best used externally for both calming and healing skin. Use leaves and flowers.

Lemon, *Citrus limon*: Improves immunity, skin health and is anti-bacterial. Use fruits.

Marjoram, sweet, *Origanum majorana*: Improves overall health, reduces joint and skin complaints, and helps calm. Use leaves.

Mint, *Mentha* spp.: Improves digestion, alleviates symptoms of colds and coughs, boosts mood, and enhances cognition. Use leaves.

Onion, *Allium* spp.: Greatly aids in general health. Use leaves and bulbs.

Passionflower, *Passiflora edulis* and *P. incarnata*: Mild sleep aid and helps ease pain and anxiety. Use fruits, leaves, and seeds.

Poppy, opium, *Papaver somniferum*: Latex used for pain relief, seeds for general health. Use seedpods and seeds.

Figure 7.6   *Aloe vera* **is an effective and fast remedy for burns, skin dryness, inflammation, rashes, and sunburn. It is the perfect plant to have around humans.** *Aloe vera* **can be grown indoor and outdoor and tolerates predation. Pictured is the yellow-flowering variety, which is the most edible.**

Rose, *Rosa* spp.: Antioxidant that improves digestion and skin health and helps calm. Use hips and petals.

Turmeric, *Curcuma longa*: Used to protect brain and improve digestion and immunity. Use roots.

Valerian, *Valeriana officinalis*: Reduces tension and restlessness. Use roots.

Trees

Alder, *Alnus* spp.: Supports digestive and respiratory health. Externally it heals skin. Use bark, leaves, and stems.

Aspen, *Populus tremuloides*: Good for digestive health, pain relief, and improves digestive functions. Use leaves and bark.

Basswood, American, *Tilia americana*: Improves digestion, lungs and skin, and helps reduce pain. Use bast, flowers, and leaves.

Bay, *Laurus nobilis* and *Umbellularia californica*: Good for digestive and respiratory health. Use leaves and nuts.

Beech, American, *Fagus grandifolia*. Expectorant and antiseptic, and good for respiratory and skin health. Use leaves, bast, and nuts.

Birch, *Betula* spp. Good for digestive and skin health. Use leaf, bast, bark and sap.

Camphor, *Cinnamomum camphora*: Some people claim it opens airways. Also used externally for joint pain. Use bark, leaves, and new shoots.

Cedar, *Cedrus* spp. Helps digestion, breathing, fevers and skins problems. Use bark, heartwood, and sprigs.

Cottonwood, *Populus fremontii*: Good for pain relief, both internally and externally. Use new shoots and bast.

Cypress, Arizona and Italian, *Cupressus arizonica* and *C. sempervirens*: Improves respiratory function and externally it is antibacterial and helps heal skin. Use cones and sprigs.

Elderberry, black and blue, *Sambucus Mexicana*, *S. nigra* and *S. nigra ssp. Caerulea*: Boosts immunity and good for colds and fevers. It also calms and helps heal skin. Use flowers, berries, leaves, and bark.
Sambucus

Elm, *Ulmus* spp.: Supports digestive health, immunity, and ease of breathing. Use bast.

Gum, blue and lemon-scented Eucalyptus, *Eucalyptus globulus* and *Corymbia citriodora*: Expectorants good for colds, coughs, stuffiness, and sore throats. Also, antiseptic and helps heal skin. Use leaves.

Fig, common, *Ficus carica*: Good for digestive health. Use fruit and leaves.

Hawthorn, *Crataegus* spp.: Used for heart health and calming. Use berries and flowers.

Loquat, *Eriobotrya japonica*: Good for colds and fevers and as a sedative. Use leaves.

Maple, *Acer* spp. (many varieties): Boosts energy and eases internal complaints. Use sap, bast, leaves, and flowers.

Mesquite *Prosopis* spp.: Good for digestion, colds and coughs, and healing skin. Use leaves, bark, seedpods, seeds, and gum.

Mulberry, *Morus* spp.: Improves digestion, respiratory, pain relief, and overall health. Use berries, leaves, and bark.

Oak, *Quercus* spp.: Provides good energy and has antiseptic qualities. Used for digestive health and healing skin. Use acorns and bast.

Pecan, *Carya illinoinensis*: Used internally for digestive and respiratory health, externally for skin health. Use nuts, leaves, and bark.

Pine, *Pinus* spp.: Good for digestive and respiratory health and externally for healing skin. Use sprigs, needles, bast, and sap.

Sassafras, *Sassafras albidum*: Supports overall health with a range of benefits. Use bark, roots, leaves, and sprigs.

Tea tree, *Melaleuca alternifolia*: Antiseptic with many external benefits. Use leaves and sprigs.

Walnut, *Juglans* spp.: Good for digestive health, heartache and helps heal skin. Use bark, leaves, and fruits.

Willow, *Salix* spp.: Brewed for pain relief. Use bark, leaves, and flowers.

Soaps and Shampoos

The material to make soap can either be found in the wild or planted and grown for the purpose. Both will need processing.

Making soap involves mixing something alkaline with something fatty/oily and then adding the accessories, such as essential oils, herbs, and colorants. Finding soap means looking for plants that contain emulsifying and surfactant properties, most of which contain saponins, a naturally occurring cleaning and foaming agent.

If making soap with conventional products, like lye and oils, do not cook food with the pots and utensils used for soap making. Use only stainless steel, ceramic or enamel pots, never aluminum, as lye will eat through it.

Nomenclature

Alkali: A surfactant that helps to remove grease and oil by lifting them from an object and keeping them in suspension. The result is an emulsion. Common alkalis for homemade soaps include baking soda (sodium bicarbonate), lye (caustic soda), and soda ash (sodium carbonate).

Base Oils: These are the oils that alter the alkali (see soapifcation), capture the essential oils, and eventually make soap. Frequently used oils include almond, coconut, grapeseed, hemp seed, olive, palm, safflower, and sunflower.

Colorants: Substances added to enhance or alter the color of soap—everything from stinging nettle to charcoal are used.

Emulsify: To combine ingredients together that do not ordinarily mix. In the case of soaps, it allows water to mix with oils and separate the dirt.

Essential Oils: The oils that provide the fragrance, health, or whatever else you want to infuse. See Fragrances above for more information.

Herbs: Dried herbs, such as lavender and sage, are commonly added to soaps for fragrance and texture.

Lye (100% sodium hydroxide): An essential ingredient in making conventional soaps. It is a surfactant. It is not an essential part of using natural soaps.

Saponins: Compounds that plants produce to protect themselves. Luckily, many of the saponins have emulsifying and surfactant properties and have been used for centuries as cleaning agents. See plant list below.

Soapifcation: The reaction of an alkali, like lye, with fat that causes it to become non-toxic and a powerful cleaning agent.

Surfactant: Literally means surface-active agent and it helps water penetrate fibers and lift dirt and oils.

GROWING A CLEANING SPONGE: THE LUFFA

In 200 days or less you can grow enough cleaning/scrubbing sponges to last 2 or 3 years. Whether washing your legs or scrubbing pans, luffas have a great degree of durability and versatility. And when they are kaput, compost them.

Luffa (*Luffa aegyptiaca* and *Luffa acutangula*) are annual vines in the cucumber family. They are voracious growers—up to 30 feet—that need plenty of room to spread. They are easy to grow in sunny areas that are frost free for about 200 days. As a rule, they are grown on trellises, because the gourds grow straight; on the ground the gourds will curve.

To harvest and prepare the sponges allow gourds to dry on the vine, withering to a flaky dirty brown. Lightly soak the gourds, peel the outer layers away, shake out the seeds out, dry for 5 days, and they are ready to use. Importantly, wash and dry them between uses, otherwise they will degrade and decompose. A sponge in use is viable for 2 weeks to 2 months. Dried sponges can be stored for years.

If you have the vine, but too many sponges already, eat the young supple fruit, which is a common food in many countries.

Figure 7.7   An abundant and effective sponge for little care or cost defines the luffa. It is practical, disposable, and fantastic at caring for and cleaning humans.

Commonly Found and Grown Soaps

The list below is comprised of the plants that can be easily found and quickly processed to create a cleansing agent. It does not include the plants that, although high in saponins and good for cleaning, demand a lot of processing. Some of these other plants include bracken fern, buckeye, horse chestnut, and yucca.

Ash, wood: Boiling ash, particularly those from hardwoods, produces a weak solution of lye that can be boiled down, added to fats/oils, and used to make soaps. Abundant.

Blue dicks, *Dipterostemon capitatus*: Corns and flowers are rubbed in water to release saponins for hand, body, and clothes cleaner. Can be abundant in open grassy fields.

Buffalo gourd, *Cucurbita foetidissima*: Fruit is cut up, simmered, and saponin rich water used to clean body and clothes.

Creosote bush, *Larrea tridentata*: Small branches of leaves are used as a body scrub. The leaves are antimicrobial. Abundant in deserts.

Elderberry, blue, *Sambucus nigra caerulea (Mexicana)*: Rub flowers between palm with water to release saponins and used for hand and body cleaner. Low suds, but effective against light grease and grime. Can be abundant.

Goosefoot, *Chenopodium californicum* and *C. murale*: Root is ground and used for hand, body and hair cleaner. Abundant in cooler areas and weather.

Ivy, *Hedera* spp.: Leaves are chopped, boiled, cooled, and the saponin rich water is used to clean clothes, hands and hair. Low suds, but great for everyday grime. Also good for fine and lightly soiled fabrics because of its softness. Abundant.

Mallow and cheeseweed, *Malva nicaeensis* and *M. parviflora*: Roots are pounded or ground in water and/or leaves are simmered; both are good for body and hair wash. Roots are small and a wash requires many plants. Leaves are high in pectin, which not only helps clean the skin but heal it too. Also good for light grease. Abundant.

Mock orange, *Philadelphus lewisii*: Flowers and leaves are agitated in water to release saponins. Used as hand and body wash.

Mullein, common, *Verbascum thapsus*: Seeds are crushed and agitated to release saponins. Can be abundant and easily cultivated.

Sage, white, *Salvia apiana*: Rub fresh leaves between palms with water for body and hair wash. Great deodorant.

Saltbush, *Atriplex lentiformis*: Crushed leaves and roots are used for washing hands and clothes. Flowers have saponins too. Abundant in Mediterranean environments.

Snowberry, common and creeping, *Symphoricarpos albus*, *S. mollis*: Smash berries in hand and water to release saponins and clean hands and body.

Soap plant, *Chlorogalum* spp. Harvest bulb for rich saponins. Fresh or dry, mash the layers in water to clean hands, body and greasy parts. Might be abundant in western facing foothill environments throughout state.

Figure 7.8   **Cheeseweed, a common weed in urban areas, can be made into a fantastic hand and body wash that also helps to protect and restore the health of skin.**

ELDERBERRY: A GIFT FROM GAIA

Elderberry (*Sambucus* spp,) should be planted in every healthy home landscape, every college campus, and every nature preserve. Elderberry deserves our protection, respect, and use. With our care and protection, elderberry provides:

- **Dye**: The fruit makes grey to purple stain, and the small stems and twigs make orange to yellow dye.
- **Fiber**: The inner bark (bast) is processed to create ornamentation, rope, and twine.
- **Medicine**: When lightly cooked the berries are antiviral and a historic remedy for colds and fevers.
- **Music**: Branches have a pithy center, which is easy to bore out and make flutes, pan pipes, rain sticks, and whistles.
- **Replication**: Elderberry is an easy plant to propagate by cuttings. The cutting should be the diameter of a pencil and 6 to 8 inches long. Potting is best during the dormant months (December thru February) and early summer months (June/July).
- **Soap**: The flower has saponins, which helps remove grease and grime. They are ground up in hand with water to create a wash.
- **Wellbeing**: The flower makes a calming tea that improves respiratory health and elevates feelings of wellbeing.

Figure 7.9   **Elderberry is a giver. While it extracts a cost—demanding care, protection and offspring—it gives more than it takes. We care for elderberry; it cares for us. That's the deal with this incredible plant.**

Resources

BOOKS

Balls, Edward K. *Early Uses of California Plants*. University of California Press 1965.

Campbell, Paul D. *Survival Skills of Native California*. Gibbs Smith. 2009.

Elpel, Thomas J. *Botany in a Day: The Patterns Method of Plant Identification*. HOPS Press, LLC. 2013.

Funk, Alicia and Kaufman, Karin. *Living Wild: Gardening, Cooking and Healing with Native Plants of California*. Flicker Press. 2013.

Garcia, Cecilia and Adams Jr., James D. *Healing with Medicinal Plants of the West*. Abedus Press. 2012.

Reid, Sara; Wishingrad, Van and McCabe, Stephen. *Plant Uses: California: Native American Uses of California Plants—Ethnobotany*. University of California Santa Cruz Arboretum. June 2009.

Slattery, John. *Southwest Medicinal Plants: Identify, Harvest, and Use 112 Wild Herbs for Health and Wellness*. Timber Press 2020.

Sweet, Muriel. *Common Edible and Useful Plants of the West*. Naturegraph Publishers, Inc. 1976.

Timbrook, Jan. *Chumash Ethnobotany: Plant Knowledge Among the Chumash People of Southern California*. Heyday Books 2007.

WEBSITES

"Desert Harvesters," a non-profit, grassroots effort at www.desertharvesters.org
"Fibershed," a nonprofit grassroots effort at www.fibershed.org
"Plants for a Future," a non-profit at www.pfaf.org

Notes

1   "About Flowers." Society of American Florists. 2018. https://safnow.org/aboutflowers/
2   Johnson, Amy. "What's the Difference in Formulation Between a Cream, a Salve, and a Body Butter?". *Plantiful Apothecary*, November 23, 2019. https://plantiful.ca/blogs/news/31383425-whats-the-difference-between-a-cream-a-salve-a-lotion-and-a-body-butter

Chapter 8

Thermal Comfort

Landscaping for thermal comfort improves the livability of urban areas. It regulates temperatures using vegetation, materials and amenities, not electricity or fossil fuels. Providing comfort naturally will reduce dependence on industrial energy, lower harmful emissions, preserve the ozone layer, protect air quality, and ultimately improve public health.

Small buildings of no more than two stories will benefit the most from these landscape practices. A landscape can reduce heating and cooling costs of a small building by an average of 25%. In hot and dry climates, the savings can be as great as 50%.[1] The cooling effect of trees cut energy costs to the point where they can pay for themselves in 6 to 8 years. In fact, the amount of greenhouse gasses reduced because of passively regulating temperatures in buildings far surpasses the amount of carbon the landscape can sequester.[2]

This chapter focuses on the landscape and its ability to affect the thermal comfort of landscapes, buildings, and communities. It covers the six strategies of cooling, the six techniques for warming, and an important discussion of maintaining landscapes and communities that provide comfort naturally.

THE UNWANTED IMPACTS OF URBAN HEAT ISLAND EFFECT (UHI)

Urban temperatures can be 5°F to 9°F warmer than nearby natural areas.[3] This is due to the density of heat absorbing surfaces, such as asphalt and concrete; the lack of cooling vegetation, such as big trees and vigorous groundcovers; and the amount of heat generating sources, such as air conditioners and cars.[4]

Urban warming gravely impacts the health of people and wildlife. It is linked to the formation of harmful air pollutants, like low-level ozone, a known respiratory reducer. It increases the chances of heat-related illnesses, such as heat exhaustion. It increases the temperature of runoff, contributing to the degradation of waterbodies. It impacts landscapes at least six miles downwind a city. It alters growing seasons.[5] And it creates a supporting feedback loop where more energy is used inside a building to maintain comfort, increasing the amount of overall heat, ultimately increasing ecological and public health hazards.

DOI: 10.4324/9781003369752-8

Nomenclature

Albedo: The reflectivity of a surface and the amount of solar energy reflected by a surface.

Conduction: The kinetic movement of heat through a material. Metal is a good conductor of heat whereas wood is not.

Convection: As air or water warms it becomes less dense and more energetic, eventually rising above the denser and cooler air or water.

Deciduous: A plant that loses its leaves once a year. Many temperate plants are winter deciduous, and some Mediterranean plants are summer deciduous.

Evapotranspiration: Moisture gains to the atmosphere through the transpiration of plants. See Transpiration below.

Evaporative Processes: Any process that releases moisture to the atmosphere, whether through evapotranspiration, evaporation or the aerosolization of water by such means as misters.

Evergreen: A plant that does not lose its leaves once a year.

Heat Index: A measure of human comfort and a combination of heat and humidity. For example, a 92°F day with humidity at 65% creates a Heat Index of 108°F, which is what a person would experience.

Heat Island: An area with higher temperatures than surrounding areas. These hotter areas typically have a lot of heat absorbing materials, such as asphalt and concrete, have little vegetation, and have sources of heat generation, such as cars and machinery.

Humidity: The amount of water vapor in the atmosphere. The amount of humidity is typically expressed as a percentage.

Insulation: A material or structure that reduces heat gain or loss by separating conducting elements. For example, a vine covering a wall will insulate stucco from solar energy.

Radiant Heat: Energy and heat transmitted as electromagnetic waves. Radiation can be heat, light or sunlight. Radiation flows in every direction, unlike convection, which flows upward.

Solar Altitude: The height of the sun. Altitude is expressed in degrees that range from 0° (horizontal) to 90° (straight up). Also called Solar Elevation.

Solar Azimuth: The position of the sun in the sky based on true north. It is expressed in degrees and for example, east would be 90°, south 180°, and west 270°.

Solar Radiation: Sunlight.

Solar Reflectance: Refer to Albedo.

Thermal Comfort: Creating environments where humans are neither too hot nor too cold. Keep reading.

Thermal Mass: The density of a material or construct that helps stabilize fluctuating temperatures. The greater the mass, the greater the stabilization. High density materials also create a lag and difference between real time and surface temperatures.

Transpiration: The process by which a plant naturally releases moisture to the atmosphere. Transpiration mostly happens on the leaves and when a plant opens its stomata to exchange gases.

Wind Compression: Compressing air to increase its speed which results in a cooling effect.

Quick Overview of Providing Comfort Naturally

Uses: Your comfort, like personal health, should be landscape 101. It works at every scale, in every garden, and within every community.

Costs: Low to high.

Difficulty: Easy to understand and design but requires time and attention.

Best Scale: Every type of landscape and all the types of people interacting with these landscapes will benefit with built-in comfort.

ERoEI: Considering that this process can be as easy as planting a tree or washing a building, the returns on energy can be fantastic—maybe the best in this entire book. The energy returns get murky and slim, however, when pipes, plastics and machinery are added to the endeavor.

Pros: Weaning ourselves off fossil fuels and cooling our warming world are modern mandates. This chapter tackles both these problems.

Cons: Using passive process to create thermal comfort will increase labor costs, the most contentious issue in landscape management. It can also greatly increase the amount of greenwaste. Please see the last section "One Big Note" for more detail.

Note: If a property or community plans to provide thermal comfort naturally, onsult the chapter on Personal/Public Health. The vegetation and maintenance practices used for thermal comfort should not produce respiratory irritants.

Figure 8.1 **Pictured is an example of growing thermal comfort and food. The grape vine dripping with fruit not only helps cool the administrative offices but also nourishes the students visiting this center. Picture taken at the Lyle Center for Regenerative Studies, Pomona, CA.**

Strategies for Landscaping for Thermal Comfort

Influencing a microclimate involves understanding the local climate. The angles of the summer and winter sun, the direction of summer and winter winds, and the humidity of summer and winter air will influence natural strategies and their success.

Unfortunately, a strategy that works well one part of the year might work too well the other part. For example, forcing wind through a breezeway may be ideal for summer, but it can make a courtyard feel like an icy cave in the winter. The strategies below will have to be prioritized. A landscape should be optimized to the season that causes the greatest discomfort, whether that be physical or financial.

The characteristics below only provide the basics of natural and passive strategies. Other publications should be consulted, some of which are listed at the end of this chapter. This chapter is divided between cooling and warming strategies.

THERMAL COMFORT[6]

Thermal comfort is defined as a temperature where people feel most comfortable. The ideal temperature varies depending on the person's amount of clothing. On average, in summer, and in light clothing, people are most comfortable in temperatures between 74°F and 78°F. In winter, with slightly heavier indoor clothing, people prefer 70°F to 73.5°F. Humidity is ideal at about 45%. Thermal comfort varies between individuals. The lower limit of comfort is 68°F and upper 78°F.

Cooling Strategies

When shading and evapotranspiration are combined an urban landscape can drop summertime temperatures by up to 9°F. When an entire community is shrouded in trees, it will be, on average, be 6°F cooler that tree-less communities.[7] And when trees are combined with vigorous ground covers, the temperature difference between the vegetation and asphalt can be as much as 25°F.

The recommendations below are ranked according to effectiveness per dollar spent. However, effectiveness is site specific. Humid areas, for example, do not benefit from evaporative processes.

A: DECREASE SOLAR EXPOSURE

Decreasing the amount of solar radiation a landscape receives and stores involves intercepting the sun, which is referred to as shading. Since the hottest sun is high in the sky, intercepting devices must be placed close to a structure or outdoor living areas to be

Figure 8.2 **A. Decrease Solar Exposure B. Increase Air Flow C. Increase Evaporation D. Increase Heat-Shedding Materials E. Increase Earth Cooling F. Insulate Surfaces**

effective. Shading devices are mostly used on the south and west sides of a building. Shading devices are divided between structures and trees.

Shade devices do more than provide thermal comfort. When houses, infrastructure, and streets are shaded, their useful life is extended. The expansion/contraction produced by temperature swings age non-living things. Without trees sunny houses need more paint, streets need more treatment, and people need more air-conditioning.[8]

Structures: Shading structures are any of the following: Arbor, awning, canopy, gazebo, pavilion, pergola, sail, solar panels, trellis, and umbrella. Fabrics and wood are the favored materials for cooling, but Plexiglas and tin are not uncommon.

Trees:[9] Some trees do a better job of shading than others. They may cast a larger shadow and/or a darker shadow, both of which improve levels of cooling. Below are the trees that typically have a large and dark shadow.

Notably, pruning plays a large role in determining the type of shade a tree casts. For example, limbing up and removing the interior branches creates a tree with great reach, but it becomes airy and its shadow light. On the other hand, tip pruning and the removing the outer gangly branches makes a more compact tree, which reduces its reach, but increases its density and the darkness its of shadow.

Good Shading Trees

There has been little quantifiable research on the amount of shading each tree produces. There is also much documented variation in the growth of a tree. Decades of field observation have been added to the scant research to create the list below.

Table 8.1:   Shading Trees

Common Name	Botanical Name	Type, Shape, Height
African fern pine	*Afrocarpus falcatus*	Evergreen, rounded, and up to 60 feet
Bishopwood	*Bischofia javanica*	Evergreen, rounded, and up to 50 feet
Incense cedar	*Calocedrus decurrens*	Evergreen, conical, and slowly to 60 feet
Golden medallion tree	*Cassia leptophylla*	Nearly evergreen, rounded, and to 25 feet
Chinese hackberry	*Celtis sinensis*	Deciduous, rounded, and to 60 feet
Carob tree	*Ceratonia siliqua*	Evergreen, rounded, and to 40 feet
Camphor tree	*Cinnamomum camphora*	Evergreen, rounded, and to 60 feet
Red flowering gum	*Corymbia ficifolia*	Evergreen, rounded, and to 45 feet
Carrot wood	*Cupaniopsis anacardioides*	Evergreen, rounded, and to 40 feet
Indian banyan	*Ficus benghalensis*	Evergreen, pyramidal to rounded, and to 100 Benjamin fig
Weeping fig	*Ficus benjamina*	Evergreen, rounded, and to 100 feet
Indian laurel fig or Chinese banyan	*Ficus microcarpa*	Evergreen, rounded, and to 40 feet
Cuban laurel	*Ficus retusa*	Evergreen, rounded, and to 60 feet
Rusty leaf fig	*Ficus rubiginosa*	Evergreen, rounded, and to 50 feet
Raywood ash	*Fraxinus angustifolia* "Raywood"	Deciduous, rounded, and to 50 feet
Arizona ash	*Fraxinus velutina*	Deciduous, pyramidal to rounded, and up to 50 feet
Chinese flame tree	*Koelreuteria bipinnata*	Deciduous, rounded, and up to 40 feet
Sweet bay	*Laurus nobilis*	Evergreen, conical to rounded, and to 40 feet
Southern magnolia	*Magnolia grandiflora*	Evergreen, rounded, and to 80 feet
Mango	*Mangifera indica*	Evergreen, rounds, and to 30 feet
Tea tree	*Melaleuca alternifolia*	Evergreen, rounded and to 25 feet
Flaxleaf paperbark	*Melaleuca linariifolia*	Evergreen, rounded, and to 30 feet
New Zealand Christmas tree	*Metrosideros excels*	Evergreen, rounded, and to 35 feet
White mulberry	*Morus alba* "Fruitless"	Deciduous, rounded, and up to 50 feet
Avocado	*Persea americana*	Evergreen, rounded, and up to 45 feet
Italian stone pine	*Pinus pinea*	Evergreen, conical to rounded, and up to 80 feet
Victorian box	*Pittosporum undulatum*	Evergreen, rounded, and to 40 feet
Long leafed yellowwood	*Podocarpus henkelii*	Evergreen, conical to rounded, and to 35 feet
Yew pine	*Podocarpus macrophyllus*	Evergreen, columnar, and up to 50 feet
Oak, coast live	*Quercus agrifolia*	Evergreen, rounded, and to 70 feet
Oak, cork	*Quercus suber*	Evergreen, rounded, and to 60 feet
Oak, southern live	*Quercus virginiana*	Evergreen, rounded, and to 80 feet
Brazilian pepper	*Schinus terebinthifolius*	Evergreen, rounded, and to 30 feet
African sumac	*Searsia lancea*	Evergreen, rounded, and to 30 feet
Coast redwood	*Sequoia sempervirens*	Evergreen, conical, and to 100 feet
American linden or basswood	*Tilia americana*	Deciduous, conical to rounded, and to 65 feet
Tipu tree	*Tipuana tipu*	Deciduous to semi-evergreen, pyramidal to rounded, and to 50 feet
Chinese elm	*Ulmus parvifolia*	Semi-deciduous, rounded, and to 50 feet
Bay or myrtlewood	*Umbellularia californica*	Evergreen, conical to rounded, and to 70 feet
Sawleaf zelkova	*Zelkova serrata*	Deciduous, rounded, and to 65 feet

Figure 8.3 **Not all trees provide enough shading. The California sycamore (*Platanus racemose*) is a good example. Its small degree of shading can be seen and felt when bicyclists stop at this rest area for relief.**

B: INCREASE AIR FLOW

Wind has a fantastic cooling effect and can drop the summertime temperatures we feel by as much as 8°F. Wind wisps warm air away from objects. Naturally, the greater the difference between the temperature of the wind and object, the greater the cooling. Wind also speeds evaporative processes, which is why people feel cooler in a breeze (our skin generally has a higher moisture content than the air).

However, there are two caveats to welcoming wind into a property:

1. **Too Hot**: The cooling effect of wind stops working above 95°F. After that wind starts adding to the heat index and feelings of discomfort. If an area regularly exceeds 95°F, blocking wind is essential in reducing discomfort.
2. **Too Cold**: The wind chill effect becomes more pronounced as temperatures drop. The cooler it gets, the more wind contributes to heat loss and discomfort. Wind should be blocked if temperatures are regularly under 76°F.

If the landscape is large, trees and walls can be used to guide wind to specific areas. Increasing airflow in urban areas often involves removing obstacles, such as large trees. The two techniques to encourage airflow are:

- *Guiding*: Whether appendages on buildings, fabric overhead, hedges, or walls near structures, wind can be guided into and out of areas and buildings. Wind behaves much like water but is fickler.
- *Funneling/Propelling*: Squeezing wind through a constricted opening will increase its velocity, which increases cooling. Compression can be a product of architecture, vegetation or walls. While wind increases in velocity at the compression point, it dramatically decreases around the feature.

*Note: Increasing air circulation is also great for landscape health. Pests, such as diseases and insects, are less of a problem in a well-ventilated landscape.

C: INCREASE EVAPORATION

Evaporation can create a large cooling effect. Naturally, the drier the environment, the better evaporative cooling works. Humid environments do not benefit from evaporative processes (although the sight of water calms and that helps cool people psychologically). Strategies for humid climes are restricted to interception, airflow, material management, and earth cooling.

Figure 8.4 **Pictured is a building that is orientated to receive the cooling summer winds. Not only are the winds compressed and sped up, but they are also guided over two small water features and through a canopy of wild grapes. Cooling is critical to this facility. Picture taken at the Audubon Center in Deb's Park, Los Angeles, CA.**

There are three sources for evaporative moisture: Plants, open sources of water, and aerosolized water:

- **Plants**: Evaporative processes favor the fast growing and water thirsty trees. Some of the best are avocado, banana, birch, catalpa, cottonwood, deciduous oak, empress tree, *Ficus*, magnolia, maple, hoja santa, mulberry, redwood, sapote, tuliptree and weeping willow.
- **Open Water**: Open sources of water include fountains, ponds, and shallow pools.
- **Aerosolizing Water**: Fountains that splash a lot of water and ponds with fountains, even if small, will increase evaporative cooling. Another effective device is misters attached to garden hoses.

D: INCREASE HEAT-SHEDDING MATERIALS

Choosing materials for cooling means managing reflectance and density.

Reflectance: Reflecting solar radiation back into the atmosphere is effective at reducing surface temperatures. Light colors rule. Roadways, driveways, walkways, and walls that are made from light colors are much cooler than dark surfaces.

However, caution is needed. There might be two unwanted impacts. First, reflecting sunlight at ground level will cause the light to bounce and that light can get under hats and sunglasses, which creates a great deal of discomfort for drivers and pedestrians. Second, that same bouncing light can slightly warm the interior of buildings and exterior walls. Aim for earth tones in urban environments, as they can provide both reflectance and comfort.

Some of the best solar reflecting materials

Density: Low density means less weight, mass, and heat storage. The heat absorbing surfaces in a landscape are the driveways, patios, retaining walls, roads, structures, and walkways. Some of the materials that can be used instead of asphalt and concrete include bricks, decomposed granite (DG), fly-ash concrete, papercrete, pavers, porous asphalt, porous concrete, turf blocks, and any wood product.

WOOD COMFORT

Of all the materials one can use in a landscape, wood is an outlier. It has two qualities that make it distinct from all other materials—both of which make wood the perfect material for managing thermal comfort.

1. Wood products have low reflectance, which means they do not disperse a large amount of the solar radiation they receive back into the environment. Radiance ranges from just 5% to 48%. Wood greatly dampens light and radiation.
2. Wood is a terrible conductor of radiation, which, of course, is a good thing, as it will not store it as heat. Wood wrapped buildings and wooden benches will help reduce discomfort on a late-summer day.

Table 8.2: Rating the Reflectance of Landscape Materials

Material	% of Reflectance[15]
Aluminum foil, bright	95
Aluminum foil, oxidized	85
Aluminum, weathered	47
Bark	23–48
Brick	23–48
Clay tiles, light colored	20
Copper, polished	75
Concrete, aged and dirty	20–30
Concrete, grey and typical	35–40
Concrete, white cement	70–80
Concrete, white cement aged	40–60
Grass	20–30
Grass, dry	32
Leaves, green	25–32
Meadow	12–30
Paint, light grey	60
Paint, white	70–75
Paint, whitewash (new)	80
Plaster, white	93
Sand, light	30–60
Soil, sandy	15–40
Water surfaces*	3–10
Wood features	5–20
Wood, pine (new)	40

*Water readily absorbs solar radiation when the sun is directly over-head. But when the sun is low, light and radiation skip off the surface and can create large and disturbing reflectance.

Figure 8.5 **The paving and seating are constructed using fly ash concrete blocks. These blocks are much lighter than conventional concrete and store considerably less heat. The color of the blocks also has a high degree of reflectance. Picture taken at the Audubon Center, Los Angeles, CA.**

E: INCREASE EARTH COOLING

The temperature 4 to 5 feet below the surface is usually much cooler than the air temperatures during summer. Temperatures vary across the nation, ranging from 55°F in the north to as high as 74°F in the south. When it is 90°F outside, even a floor at 74°F feels fantastic. Below are three common methods to increase earth cooling.

Sub-Grade: Sub-grade construction means putting a structure underground. The structure may be partially or wholly underground. Despite being the most effective method for passively regulating the temperatures of a building, it is not widely used. Building down into the earth is far more expensive than building on top. Refer to the Energy chapter and "Earth Energy" for greater detail.

Berm: A berm is a compacted pile of earth that is usually vegetated. Berms are built up against structures to increase thermal protection. They are much less expensive than sub-grade construction. Mounded soil can also be used to create outdoor rooms. But berms do more than just provide thermal mass, they also buffer the wind and increase the amount of planted area.

Slab On Grade: The cooling of concrete and other hard surfaces is sped by placing the dense surface directly on grade (highly compacted soil). Typical construction calls for

Figure 8.6   **These benches provide refuge to many on a hot day. Their mass is imbedded into the earth and cooler as a consequence. They are also shaded from the afternoon sun and good tree maintenance ensures ample airflow.**

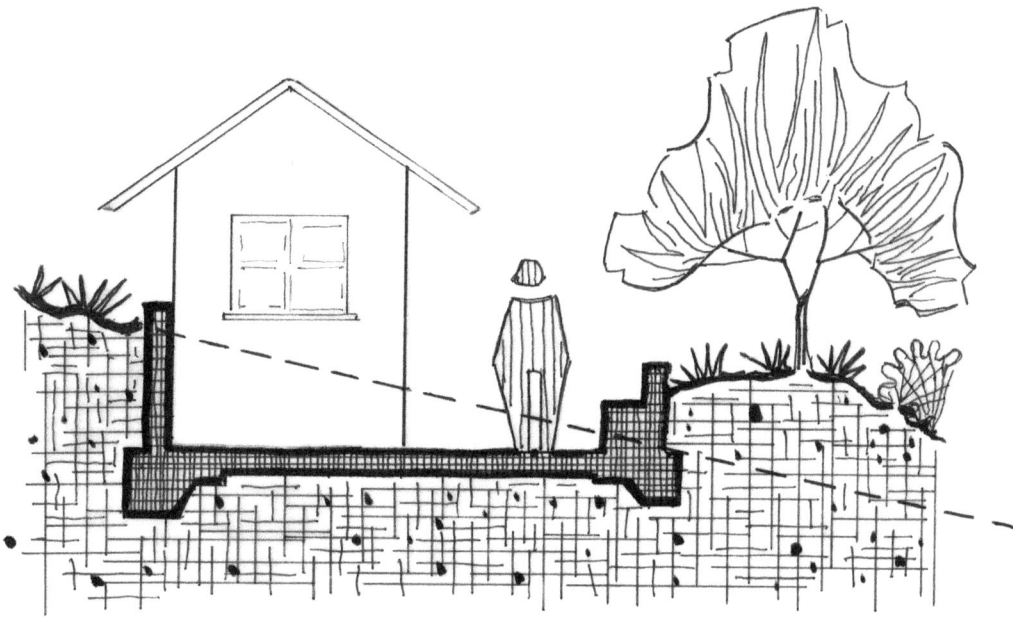

Figure 8.7   A cross section illustrating sub-grade, slab on grade, and berms to increase thermal massing and help regulate extreme temperatures.

concrete to be poured on a bed of sand and/or gravel. Slab on grade, however, is only used for low impact surfaces, like driveways, walkways and low use parking lots. Putting bricks and pavers directly on grade is much less resource and time consuming than placing them on sand.

F: INSULATE SURFACES

Berms, greenroofs, living walls, and even vines scrambling up a wall are all excellent at protecting structures from heat gain and loss. These techniques work because of shading, thermal mass, and evapotranspiration.

Two caveats about greenroofs and living walls. First, while living surfaces can feed pollinators, grow food, clean and slow runoff, provide recreation, and increase people's feelings of wellbeing, they are also expensive. They can add up to 10% to 15% of the total building costs, and demand near weekly maintenance.[10] And second, irrigating a feature attached to a building can moisten the building, reducing its longevity and the health of its occupants. A dry building is a healthy building.

WHERE TURF MAKES ECOLOGICAL SENSE[11]

When it comes to choosing the coolest type of landscape for dry urban areas a study in Israel helps lead the way. Researchers tested six types of land cover and evaluated each on cooling effect and water use.

The strategy with the greatest cooling was shaded turf. A synergic effect was created when turf was either shaded by shade cloth or trees and both methods were able to drop a 93°F afternoon to 88°F. Shade cloth alone actually raised the air temperature by 1°F to 2°F. Turf without shading was one of the poorest investments; it had marginal cooling effect but demanded the most amounts of water and maintenance. Trees without ground cover were able to drop temperatures 3°F to 4°F.

Cooling Communities

Below is a list of community-wide cooling strategies that can, and have been adapted by city and county planning agencies across the nation. These strategies are listed in order of the amount of cooling effect per dollar spent.

- **Light Reflecting Roofs**: There are many light and heat reflecting paints and membranes, like vinyl, that can be applied to roofs. These relatively inexpensive treatments can reduce the energy needed to cool a building by as much as 10%. The Cool Roofs Rating Council (CRRC) can help with picking the best coating and surface.

Figure 8.8   **The exterior of this small train station was designed to accommodate a vine growing on its western wall. The Boston ivy (*Parthenocissus tricuspidata*) pictured here helps regulate both indoor and outdoor temperatures. The vine is only allowed to grow to 8 feet tall, which makes maintenance easier and less expensive.**

- **Light Colored Concrete**: When compared to asphalt, light colored concrete can reflect up to 50% more solar radiation.
- **Light Colored Surfaces**: Whether the side of a building or walkway, the color of pavers or decomposed granite, almost every material used in a landscape comes in a variety of colors. Naturally, the lighter these surfaces, the more reflectance and cooling they are.
- **Curbside/Medium/Parkway Planting**: Even if light colored, dense surfaces, like sidewalks and roadways, absorb a lot of heat. Shading these surfaces is the surest way to reduce their temperatures. Planting canopy trees, those umbrella-like trees, in parkways and street mediums will lower temperatures.
- **Open Space Planting**: Plants, especially trees, are much better at cooling than dirt. A community shrouded and surrounded by trees can be up to 6°F cooler than comparable, but tree-less communities. Keeping a community vegetated will not only help regulate temperatures, but also improve air quality, clean and slow run-off, and increase property values.
- **Vegetated Vertical Surfaces**: Per unit of cooling, vegetated building surfaces are by far the most expensive. However, greenroofs and living walls can provide other services that may be in short supply in urban areas. Besides cooling, these vegetated surfaces can also help clean the air, slow and screen runoff, provide food,

Figure 8.9 **For concrete and car exhaust, humans are rewarded with intense shading, stormwater diffusion, and pollutant removal. The Moreton Bay fig (*Ficus macrophylla*) pictured here has been providing thermal comfort along this busy street for over 148 years. Picture taken in Orange, CA.**

increase pollinators, and boost feelings of wellbeing. An inexpensive alternative to built structures is to let vines, such as climbing fig and Virginia creeper, grow up a wall.

Warming Strategies

The recommendations below are ranked according to effectiveness. However, effectiveness is site specific. For instance, snow is an excellent insulator and if used as such, it should be shaded (not welcome the sun).

A. DECREASE WIND

Reducing wind speed and consequently wind chill will reduce the amount of energy needed to heat a building by 7% to 25%.[12] It also greatly impacts the temperatures we feel in a landscape. A 10mph wind in 40F will make it feel like 32F, in 20F it will feel like 9F. Blocking winds can help make the outdoors livable in winter. The goal is to put obstacles as possible in the path of wind and confine the air, allowing it to warm up.

Below are some of the types of windbreaks. Naturally, the taller and thicker the windbreak, the larger the wind shadow and protective effect.

Berms (earth): Earth can be piled up against a structure to protect and insulate it. Berms and flowing landforms can be constructed to disrupt airflow in the landscape.

Fences: Less expensive than a wall and semi-permeable to airflow, fences disrupt airflow differently than walls. Small fences are often installed along roads to prevent the build-up of snow on roads. The permeable nature of fences means that it impacts wind and its direction, speed and turbulence differently than a solid wall.

Shrubs and Trees: Densely growing and preferably evergreen trees and large shrubs are used to create windbreaks. Some of the best shrubs are bay, elderberry, hawthorn, juniper, lilac, Manzanita, privet, *Rhus*, silverberry, *Viburnum*, willow, and witch hazel. The best trees include arborvitae, blue gum *Eucalyptus*, carob, cedar, cypress, bottlebrush, *Ficus* (some), *Feijoa*, fir, hemlock, huckleberry, oak (live), olive, peppertree, pine (some), *Podocarpus*, redwood, spruce, *Syzygium*, tea tree and yew.

Walls: Impermeable walls are excellent at helping to protect outdoor living areas and structures from wind chill. Because walls lack depth, their effect of wind speed is short lived, and walls should be used close to the things they are trying to shield.

B. INCREASE SOLAR GAIN

Increasing the amount of sunlight that strikes warming surfaces, such as walkways and walls, is the second most effective warming strategy. If the property is wooded, this strategy entails a lot of vegetation management. Below are the most common ways to increase solar gain.

Deciduous Vegetation: Deciduous plants are used where solar penetration is needed in winter, but not summer. However, deciduous plants can still deflect/absorb anywhere between 20% and 70% of the sunlight.[13] Some of the plants used for summer shading and winter warming are dogwood, elm, flame tree, gingko, grape, liquidambar, oak, pomegranate, *Prunus*, redbud, and sycamore.

Sized Shading: Shading devices such as awnings, eaves, louvers and overhangs should be precisely sized to block the summer sun but allow the lower winter sun full penetration into a building and outdoor living spaces.

Reflecting Wall: Generally installed on the north side of buildings, these walls reflect the southern sun back to the landscape and building. Reflecting walls are made from dense materials, such as concrete, and light colors to increase reflectiveness. They are not wooden.

Sunroom: A sunroom is a transparent structure that allows full sun penetration but blocks wind and its cooling effect. The heat generated in sunrooms often warms the structure it is attached to.

Figure 8.10 **The winter sun can penetrate this structure and provide natural warming. The grape planted on the pergola outside the structure blocks the hot summer sun, but the deciduous vine and width of pergola allows for solar penetration in winter. Picture taken at the Lyle Center for Regenerative Studies, Pomona, CA.**

C. DARK AND DENSE MATERIALS

Some materials are naturally better than others at absorbing the sun's energy and storing it as heat. These materials are dense and dark. Listed in order of ability to absorb and store heat, the best materials are steel (by far), marble, wet soil, granite, concrete, water, brick, and adobe.[14]

Asphalt absorbs a lot of energy and will warm, but it does not store the heat for long as concrete. It lacks the density. Asphalt is considerably less expensive to install, sometimes by as much as 45%, despite having greater maintenance costs and a shorter life span. Asphalt is ideal in cold regions where it lasts longer and can melt snow faster than concrete.

Some of the materials that absorb solar radiation and have low reflectance.

Table 8.3: Rating Landscape Materials on Radiation Absorption and Reflectance

Material	% of Reflectance[16]
Asphalt, new	5–10
Asphalt, old	10–15
Blacktop	10–15
Bluestone (sandstone)	18
Brick	23–48
Concrete, aged and dirty*	20–30
Concrete, textured*	25
Concrete, dark colored (new)*	10–30
Granite*	45
Iron, galvanized and aged*	10–20
Marble, polished*	40–50
Paint, black gloss	10
Paint, dark grey	30
Paper, black tar	7
Soil, dark	7–10
Slate, dark*	7–10
Steel, aged*	5–20

* Denotes a high-density, heat-holding material.

D. INSULATE

There are two aspects of providing insulation in a landscape. The first is to insulate heat-absorbing materials from heat-robbing materials. The second is to insulate a structure with vegetation.

First, materials that store heat need protection from materials that rob it. In a landscape the two biggest heat sinks are the atmosphere and earth. Heat absorbing surfaces, such as driveways, steps and walls, are separated from the earth with a generous layer of sand or gravel, which have air space and do not readily transfer heat and radiation. Protecting the surface from the atmosphere and wind involves using transparent surfaces, such as glass, Plexiglas or a sunroom.

Second, vegetated berms, greenroofs, living walls, and vines scrambling up a wall are good at protecting structures from heat loss. Vegetated surfaces provide these benefits by shielding a structure from wind chill and providing thermal mass.

E. INCREASE EARTH WARMING

Whether summer or winter, using the mass of earth is fantastic at regulating extreme temperatures. The trick is to root a structure and/or outdoor living area in the denser soil 2 to 4 feet down. Please refer to the discussion "Increase Earth Cooling" above.

F. COMPOST

While it takes tens of thousands of pounds of compost to create the heat needed to radiate into a building and warm it, stacking any amount of mulch against a building will shield and protect it, ultimately reducing the building's heat loss through insulation. Notably, there must be a waterproof barrier, such as pond liner, between the building and pile.

For a deeper discussion on the many ways to utilize compost for heat, please refer to the Energy and Landscape Materials chapters.

One Big Note

Creating thermal comfort naturally isn't as easy as flipping a switch like it is with industrial energies—it demands our time and attention. The lower half of the nation's urban areas would benefit from cooling strategies, which means an increase in vegetation. These areas will need to factor in more labor time and higher expenditures.

Properly maintaining vegetation and surfaces are as important to urban cooling as they are to public health.

Vegetation Maintenance

Whether it is fallen fruit discoloring light reflecting surfaces, roots uplifting walkways, or limbs prying off roof tiles, urban vegetation demands care. Cleaning, pruning, regeneration, creating avenues for greenwaste, and irrigation are vital to a healthy urban forest. They are briefly discussed below.

PRUNING

Urban vegetation needs pruning, but not always at the frequency and intensity as commonly practiced. Below is a list of reasons for pruning. If a plant is not on this list, it may not require pruning. Many experts consider pruning without a good reason horti-torture, not horticulture.

- **Safety**: Prune any part of a plant that poses a risk to humans. Removing roots lift sidewalks and pruning branches at eye level as two common examples. Any thorny or poisonous plants 2 feet on either side of a walkway should also be pruned, if not removed.
- **Infrastructure**: Cable, electric, gas, sewage, storm drain, and water pipes—so much of our urban infrastructure runs underground. Protecting these essential arteries means pruning and corralling roots and shoots. There are specific methods for pruning aerial hazards and underground infrastructure. Please, consult and/or contact your local utility provider for details.
- **Plant Health**: The continual removal of the Three Ds: Dead, damaged and diseased wood is the single best use of limited time. These weak and susceptible parts of a plant are the entry points for plant killing pathogens and the kindling for fire. The next priorities for pruning are the removal of parasites, like mistletoe, limbs rubbing together, and strength robbing suckers and watersprouts.

REGENERATION

Urban forests are not planted but regenerated. Removal and replacement are ongoing.

Figure 8.11   This downtown business district knows the value of urban trees and community cooling. Instead of concrete, they use permeable pavers around their street trees. Pavers allow the soil to exchange its gases, rainwater to percolate into the soil, and rapid repair of uplifted sidewalks. Picture taken in downtown Tustin, CA.

There comes a point in every urban plant's life where the economic and/or ecological costs of upkeep exceed the environmental benefit. It is at this point, and hopefully no further, when a plant is replaced. Trees and large shrubs, for instance, need to be replaced every 10 years in alleyways. Beyond 10 years and the road will begin to buckle, walls lean, and powerline pruning becomes more frequent and costly. Removing and replanting in urban areas requires courage because backlash from removing vegetation is common.

Life Expectancy Generalized

Table 8.4:   Generalized Life Expectancy of Landscape Plants

Plant Group	Expected Life
Large Trees	30–120 years
Medium Trees	20–60 years
Street Trees	8–20 years
Large Shrubs	10–25 years
Small Shrubs	5–15 years
Vines	5–20 (except asexual vines which readily reproduce by root divisions)
Perennials	2–6 years
Annuals	Every Year

AVENUES FOR GREENWASTE

A community committed to urban cooling through vegetation must be equally committed to efficient greenwaste management. Increasing the number of trees will increase the volume of greenwaste. Currently there are four ways to manage greenwaste in urbanized areas: Burning, mulching, landfilling and bioutilization.

*Note: A smart land manager will use plants that have value at the end of their life. For instance, this value might be in craft items, energy potential, or logs for lumber.

Burning is the quickest and least expensive method, but legal in only the rural and agricultural regions. Burning used to be the most common method of disposal, but air quality and public health mandates dictate a wiser approach in dense urban areas.

Mulching/Composting is an inexpensive method but requires the greatest amount of time and space. The shredding of greenwaste creates mulch, compost and humus. These products provide weed suppression, water conservation, surface cooling, plant nutrients, and food for soil ecology. A site's ability to compost hinges on its ability to use the products; there is little point to composting if it will not be used.

Landfilling greenwaste is the most expensive method of removal. Landfilling involves hauling, the cost of which is related to weight. Drying vegetation before hauling can reduce costs. The biggest benefit of landfilling greenwaste is carbon storage. All the other techniques mentioned here involve the decomposition of biomass and the quick release of its stored carbon.

Bioutilization makes use of vegetative waste. The greenwaste can be turned into energy, mulch and/or product. Mulch is discussed above and briefly discussed below are energy and product.

Energy: Greenwaste can be used to generate electricity. The two most common methods of generation are direct use, which is electricity from incineration, and indirect use, which captures the methane from landfills. Please refer to the Energy chapter for the specifics of each.

Product: Craft items, pallets, paper and landscape materials are some of the products greenwaste can be turned into. Creating these products is something that should be done onsite. Repurposing greenwaste at the municipal level is problematic because the sources are varied and greenwaste will have many types of foreign objects (metals and plastics), pathogens, plants, weeds, and miscellaneous pollutants (pesticides, salts, trash).

WATERING

Tree roots will surface and cause damage when they are over or under watered. Too much water and the soil becomes oxygen deprived, driving the roots upward for air. Too little water and roots will come to the surface to compete with surrounding plants.

The trick is to keep the first 2 inches of the soil dry and the soil 6 inches to 12 inches down slightly moist. Deep and infrequent irrigation is the technique to ensure proper soil moisture.

Figure 8.12   Moss and surfacing roots are sure signs of too much water and too little oxygen, as seen with this peppermint tree (*Agonis flexuosa*) along a sidewalk.

Surface Maintenance

Managing materials and surfaces is ongoing in urban landscapes. Everything loses efficiency over time. The goal of surface maintenance is to maintain the design efficiencies and the tangible benefits these surfaces provide.

CLEAN

Light reflecting surfaces, such as concrete walkways and walls, will get muddy colored over time and consequently become better at absorbing heat (opposed to shedding it). Power washing is needed and often a yearly task. Permeable surfaces, like bricks and pavers, require vigorous sweeping and vacuuming. And transparent surfaces, such as skylights, need cleaning annually.

SURFACE TREATMENTS

Every other year expect to touch up and/or resurface a light reflecting roofs and walls. Decomposed granite, a heat reflecting walking surface, will need muddy areas scrapped off. Surfaces on the east and north sides of structure may need treatment for algae.

WOOD

If wood is natural (not treated), then it will need replacing every 10 to 15 years. If wood is treated, then the wood will need painting/preserving every 2 to 5 years.

Resources

BOOKS

Brown, Robert D. and Gillespie, Terry J. *Microclimate Landscape Design: Creating Thermal Comfort and Energy Efficiency*. John Wiley & Sons, Inc. 1995.

Cotterell, Janet and Dadeby, Adam. *The Passivhaus Handbook: A Practical Guide to Constructing and Retrofitting Buildings for Ultra-Low Energy Performance*. Green Books. 2012.

Labs, Kenneth and Watson, Donald. *Climatic Building Design: Energy-Efficient Building Principles and Practice*. McGraw-Hill Book Company. 1983.

Lyle, John Tillman. *Regenerative Design for Sustainable Development*. John Wiley & Sons, Inc. 1994.

WEBSITES

Climate Central. Can be viewed at https://www.climatecentral.org/

Green Infrastructure, American Society of Landscape Architects (ASLA). Can be viewed at https://www.asla .org/

U.S. Green Building Council. Can be viewed at https://www.usgbc.org/

Notes

1 Simpson, J. R. "Improved Estimates of Tree-Shade Effects on Residential Energy Use." *Energy and Buildings*, vol. 34 no. 10, November 2002, pp. 1067–1076. https://www.sciencedirect.com/science/article/abs/pii/S0378778802000282

Ko, Yekang. "Trees and Vegetation for Residential Energy Conservation: A Critical Review for Evidence-Based Urban Greening in North America." *Urban Forestry & Urban Greening*, vol. 34 August 2018, pp. 318–335. https://www.sciencedirect.com/science/article/abs/pii/S1618866717306325

2 Jo, Hyun-Kil and McPherson, Gregory. "Carbon Storage and Flux in Urban Residential Greenspace." *Journal of Environmental Management*, vol. 34 June 14, 1995, pp. 109–133.

3 Lyle, John Tillman. *Regenerative Design for Sustainable Development*. John Wiley & Sons, Inc. 1994.

4 Cook-Anderson, Gretchen. "Urban Heat Islands Make Cities Greener." *NASA*. June 6, 2004. https://www.nasa.gov/centers/goddard/news/topstory/2004/0801uhigreen.html

5 Cook-Anderson, Gretchen. "Urban Heat Islands Make Cities Greener." *NASA*. June 6, 2004. https://www.nasa.gov/centers/goddard/news/topstory/2004/0801uhigreen.html

6 Watson, Donald and Labs, Kenneth. *Climatic Building Design: Energy-Efficient Building Principles and Practice*. McGraw-Hill Book Company. 1983.

7 "Landscaping for Energy Efficiency". South Carolina Energy Office. No date. http://www.builditsolar .com/Projects/Cooling/Shading/EB%20Landscaping%20for%20energy%20efficiency.pdf

8 McPherson, E. G. and Muchnick. J. "Effects of Street Tree Shade on Asphalt Concrete Pavement Performance." *Journal of Arboriculture*, vol. 31, no. 6, November 2005, pp. 303–310. https://www.fs.usda .gov/research/treesearch/46009

9 "SelecTree: A Tree Selection." Urban Forest Ecosystems Institute. https://selectree.calpoly.edu/

Kourik, Robert. *Designing and Maintaining Your Edible Landscape Naturally*. Metamorphic Press. 1986.

Spalding, George H. "Some Outstanding Shade Trees for Southern California."

Lasca Leaves, vol. 21, 1971, pp. 64–70. https://www.arboretum.org/some-outstanding-shade-trees -for-southern-california/

Street Trees Recommended for Southern California, 2nd ed. Street Tree Seminar, Inc. 2000.

Sunset Western Garden Book. Sunset Publishing Corporation. 2001.

And lastly, my experience played a part of the selection of these trees. Having waited at bus stops throughout the state I have gotten good at identifying effective shade trees.

10 Ieszic Formeller. *A Healing Home: The Application of Therapeutic Landscape Design Theory to the Residential Setting*. Master's Thesis accepted by California State Polytechnic University Pomona, Department of Landscape Architecture. June 2010.

11 Shashua-Bar, Limor, et al. "The Cooling Efficiency of Urban Landscape Strategies in a Hot Dry Climate." *Desert Architecture and Urban Planning*. The Jacob Blaustein Institutes for Desert Research, Ben-Gurion University of Negev, Israel, vol 92, no 3–4, May 27, 2009, pp. 179–186. http://dx.doi.org/10.1016/j.landurbplan.2009.04.005

12 *All About Trees*. Ortho Books. Edited by Barbara Ferguson. 1982. They said houses in windy areas that have successfully block the wind reduce energy needs by 25%.

Coder, Rim D. *Identified Benefits of Community Trees and Forests.* The University of Georgia Cooperative Extension Service Forest Resources. Publication FOR96-39. October 1996. https://nfs.unl.edu/documents/communityforestry/coderbenefitsofcommtrees.pdf This study stated that a 50% reduction of wind speed reduced heating energy by 7%.

13 Kourik, Robert. *Designing and Maintaining Your Edible Landscape Naturally.* Metamorphic Press. 1986. p. 75.

14 Watson, Donald and Labs, Kenneth. *Climatic Building Design: Energy–Efficient Building Principles and Practice.* McGraw-Hill Book Company. 1983.

15 Donald Watson and Kenneth Labs. *Climatic Building Design: Energy–Efficient Building Principles and Practice.* McGraw-Hill Book Company. 1983.
and
"Albedo: A Measure of Pavement Surface Reflectance." American Concrete Pavement Association. June 2002. http://overlays.acpa.org/Downloads/RT/RT3.05.pdf

16 Watson and Labs. *Climatic Building Design.* McGraw-Hill. 1983.
"Albedo: A Measure of Pavement Surface Reflectance." American Concrete Pavement Association. June 2002. http://overlays.acpa.org/Downloads/RT/RT3.05.pdf

Chapter 9

Timber

We are all encouraged to shop locally, reduce our energy consumption, and use the power of plants to help cool our neighborhood to create more livable cities. Growing urban trees for timber can do all those things and more.

There are about 69 million acres of urbanized forest in the U.S. This forest resides within public installations and facilities, in residential and recreational areas, and along roads and other transportation corridors. The US Forest Service claims that approximately 3.8 million board feet of timber could be produced from urban tree waste each year, which is about 30% of the nation's hardwood lumber demand.[1] What's more, studies say that timber trees grow faster in urban environments than they do in the wild, the reasons being the richer soil, more water, and higher amounts of carbon dioxide and nitrogen oxides. In Southern California, for instance, trees in public parks had upwards of 37% more wood at the same age than their counterparts in the wild.[2]

Despite the benefits and potential of urban timber production, the practice is not widespread in the U.S. Included in this chapter are an overview of the two types of timber production (salvage and intentional) and the strategies to make intentional production a success, which includes design guidelines, capital requirements, a list of high value timber trees, and maintenance considerations.

Two Type of Timber Production

SALVAGE HARVEST

Salvage harvest relies on trees that were not originally planned for the purpose of lumber. The trees are large, straight, and composed of valuable wood, but need to be removed because they are dying, their roots or branches have become a hazard, they have become too expensive to maintain, or they need to be removed to make way for another type of land use.

DOI: 10.4324/9781003369752-9

Figure 9.1   **Angel City Lumber generates many timber products from naturally felled trees across Los Angeles County. Pictured is their sorting yard in Boyle Heights, CA.**

Salvage harvest is an unpredictable and expensive process. There is no control over the quality of the wood, how much wood is available, or the timing of harvest, which is dictated by the company contracted to remove the tree. Consequently, salvage harvest produces expensive wood that is sold for artisan, boutique, and monied green-building pursuits. There is a growing appetite for this kind of material, as the burgeoning businesses across the U.S. have proven. However, the price of salvaged timber dissuades everyday general use. It is a niche and will likely remain so until prices can be lowered, which depends on making quality, quantity, and timing of harvest more predictable and reliable.

INTENTIONAL TIMBER PRODUCTION

Intentional production is the practice of planting high value trees in urban areas for the purpose of harvest. Intentionally planting and harvesting timber trees solves many of the problems with salvage harvest. It helps create predictability in the timing, quality, and quantity of harvest—it increases efficiency, ultimately reducing the price of urban timber. While the practice in currently uncommon, intentional harvest was widespread in Colonial U.S. and post-WW2 Europe.

Planting trees designed to be harvested for timber would produce modest and substantial changes. The modest part is that low value trees would be replaced with high value ones, and maintenance chores are focused on reducing injury to trees and maintaining straight trunks. The substantial part is the cyclic removal of trees while they are still alive and healthy. A viable timber program needs healthy trees for good timber. Urban residents aren't used to this.

Figure 9.2   **Intentionally planting trees for timber in urban areas enhances predictability of harvest, ensures higher quality, and ultimately leads to lower prices and a greater range of products. Pictured is dimensional lumber, which is a hard product to sell from salvage timber because of the low reliability and high price.**

This chapter only focuses on intentional timber production and highlights the best types of landscapes, the ideal trees, and the maintenance attributes that ensure high quality wood.

Quick Overview Urban Timber

Uses: Most of the lumber urban areas produce is considered boutique wood, which is used for artwork, cabinetry, decoration, and outdoor furniture. Load-bearing wood must meet higher standards, and the variability of current urban timber makes these standards cost and scale prohibitive. Low quality timber is used for landscape materials (please refer to that chapter).

Costs: High.

Difficulty: There are two hurdles. First, proper maintenance is essential (read further on). Second, felling long useable trunks from within congested urban areas is problematic, dangerous, and expensive.

Best Scale: As you will read later, an economically sustainable timber program would involve no less than 700 managed acres. The best landscapes are huge regional parks managed by large entities, such as a state, county or city. Residential properties can produce valuable timber, but the economic and energy costs of harvest may be cost prohibitive.

ERoEI: Even with the greater transportation costs of imported lumber, it is hard to beat industrial timber production in terms of imbedded energy and per unit costs. Urban areas lack economies of scale and the energy investments per board foot are higher. However, when the other benefits of urban forests are factored, intentional urban production makes economic and energetic sense. These other benefits include carbon sequestration, public health, and urban cooling.[3]

Pros: Conservation of resources, carbon sequestration, public health, revenue, bird habitat, and diversion of waste are some of the many benefits of growing urban timber.

Cons: The biggest hurdle is the difficulty of selling the public on producing public goods on public land. Watching healthy trees being farmed will upset some people.

Nomenclature

A/D: Air dried lumber is simply stacked, protected from the elements, and allowed to naturally dry. The process can take up to two years.

Figure 9.3 **Urban lumber mills offer a wide range of products from a wide variety of trees. From hardwood to softwood, slabs to dimensional lumber, these companies work hard to repurpose and sell urban biomass. Pictured is the showroom of Street Tree Revival in Anaheim, CA.**

Board Feet (bf): A unit of wood measuring 1" thick by 12" by 12": 144 cubic inches.

Bucking: The process of cutting a felled tree into logs.

Butt End: The part of a sawlog with the greatest diameter. The butt end is the area closest to the base (flare).

Chip-N-Saw (CNS): A type of lumber produced from medium-sized pine trees. The CNS method grinds away the outermost layer of wood, and then saws lumber from the interior wood of the log beam, making 2" × 4" boards and other lumber.

Commons: Second grade of hardwoods. Used for smaller projects, like cabinets.

Compression Wood: When a trunk or branch leans the cells of the wood on the underside becomes compressed. The cells on the topside become elongated, called tension wood.

Cull: A tree or log that has a defect that makes it unusable. Culls may be caused by natural defects, such as rotten wood or knots, or by mistakes made when the log was harvested or prepared for shipment.

Diameter at Breast Height (dbh): The diameter of a tree 4½ feet above the ground on the uphill side of the tree. Dbh is one of the two measures used to determine the amount of board feet in a log, the other is merchantable height.

Dimensional Lumber: Lumber that is finished/planed and cut to standardized width and depth specified in inches.

Finished Size: The size of a board after finishing, which is typically 3 mm less than the nominal size on all machined sides.

Flare: The area where the trunk flares and meets the roots or buttress. Flares are typically bulbous and easy to spot.

Hardwood: As opposed to softwoods, hardwoods are generally more valuable. These trees have broad, flat, or scalloped leaves. Hardwoods have their seeds contained within a nut, fruit, berry or other outer casing (angiosperms). This group of trees can be deciduous or evergreen.

Insect Defects: Damage caused by wood eating insects and/or the pathogens they carry. Insect defects reduce the quality of lumber.

Intentional Harvest: The practice of planting high value trees in urban areas for future harvest and timber products. Intentional timber production can improve quantity and quality and increase sellable yield by as much as 74%, diverting 41% more carbon than a salvage timber program.[4]

Kiln Dried: Lumber dried in a kiln. Most commercial lumber is kiln dried because it maintains quality and speeds the drying process. The moisture content is usually 12%.

Log Rule/Log Scale: Tables of estimated yields from logs or trees. There are approximately 95 accepted log rules. All rules are based on mathematical formulas.

Lumber: A term for timber that has been cut into boards.

MBF: Abbreviation for 1,000 Board Feet. The MBF is standard for volume estimates when using Board Foot as a measure or a Log Scale.

Merchantable Height: The merchantable height of a tree describes the maximum height at which point the tree is usable for lumber, pulpwood or other forest products.

Milled Products: Wood products made from the log of a tree (sawlog) in a sawmill.

Nominal Size: The original size of a board before machining, such as a 2" × 4". Also see "finished size".

Non-Milled Products: Non-milled products demand less processing and are of lesser value. Some of the products are firewood, fiber, fuel, mulch, poles, pulpwood and sawdust. See Landscape Materials.

Pallet Wood: Lumber used to manufacture wood pallets. This wood is low quality.

Plylog or Peeler Log: A log that is large enough and of suitable quality to manufacture veneer sheets or plywood. Good plylogs are straight, relatively free of knots, and have good quality. Plylogs have a value equal to or greater than sawlogs.

Pole Log: A tree that has a specific taper, straightness, diameter, and overall form that makes it suitable for utility poles. Pine trees are often used as pole logs. They are expensive because handling a 30-foot to 40-foot pole is difficult.

Pulpwood: A log used to manufacture paper, absorbent pulp, cardboard, fiberboard and other wood fiber-based products. The trees that produce pulpwood are of the lowest quality.

Quality: The quality of wood affects its price. The qualities that are examined in the field are straightness, no significant signs of damage or injury, no materials imbedded in the tree, and no signs of dying and/or decaying wood.

Rough Sawn: Lumber sawn on a bandmill or large circular saw to approximate dimensions.

Salvage Harvest: Wood products harvested from trees that are naturally removed in urban areas.

Sawlog or Sawtimber: A log or tree that is large enough and of suitable quality to be sawn into lumber. Generally, a sawlog must be straight, relatively free from knots, and have sound wood. It is also typically no less than 6 feet in length. Large diameter trunks and hardwoods are the most valuable.

Sawer/Sawyer: The person milling the wood.

Silviculture: The art, science and practice of establishing, cultivating and tending to a forest.

Small End: The part of a sawlog with the smallest diameter. The small end is the part closest to the scaffolding branches.

Snag: A dead, decaying and standing tree. Snags are good for habitat but make poor lumber. Snags might be used for non-milled products.

Softwood: Typically, softwoods have needles or scale-like leaflets. They are mostly conifers. Most are evergreen, but some, like larch, are deciduous. Seeds are usually protected in cones. Softwoods are typically lighter than hardwoods and are generally less valuable. Most softwoods are gymnosperms.

Sustained Yield: Management of a forest to produce relatively consistent quantities of wood product, revenue dollars, or some other measurable result over a long period of time.

Sweep: The curve along a trunk or branch. A large sweep will increase the amount of compression and tension wood, while also decreasing its quality and value.

Tension Wood: When a trunk or branch leans the cells of the wood on the topside become elongated, called tension wood. The cells on the bottom side tighten, called compression wood.

Timber: Trees are considered a source of wood for lumber and related uses.

Trunk: The part of a tree that starts at the flare and ends at the highest scaffolding branches. The trunk is sometimes referred to as the Useable Length. Also referred to as "bole".

Urban Artifacts: Urban trees can have items imbedded into the trunk, which makes milling them dangerous and expensive. Artifacts might include fencing, nails, rocks, signs, and toys.

Usable Length: The part of a trunk or branch that has value for lumber.

Design Strategies

Growing timber is less of a technical challenge than a social challenge. Simply growing high value timber trees is not enough for a successful program; the infrastructure and social structures must be in place to ensure that the wood gets harvested, milled, and bought.

Large lumber companies can sell wood cheaper than local organizations because they possess economies-of-scale. They manage hundreds of thousands of trees, not thousands. For an organization to have an economically and ecologically sustainable program there must be a minimum number of trees, trucks, mills, drying racks, and buyers.

The goal of a sustainable enterprise is to keep the mill and retail operation running at near capacity; steady work lowers per unit costs. For a small urban mill approximately 458 trees need to be milled a year. Assuming a 15-year harvest, a total of 6,870 trees are required for a sustainable venture. A 15-year-old tree in urban areas will produce more than 150+ board feet (bf) of lumber and at $10 per board feet retail, each tree is worth more than $1,500, although some trees, like walnut, sell for much more (and others, like pine, for less).

WHAT IF...

What if the Parks Department of Orange County, CA, pursued timber production?

Orange County manages 15 regional parks, seven of which are perfect for timber production—they are large, open and grassy. Total harvestable acreage is 1,666. If the county

Figure 9.4 **The trees naturally removed from this park are not turned into lumber. However, if trees were intentionally planted and maintained for lumber, this park, which has 163 harvestable acres and has approximately 1,715 trees, would be an ideal landscape for timber production. Picture taken at Mile Square Park, Fountain Valley, CA.**

converted to intentional production, the operation would manage 17,528 trees, harvest 879 trees a year, handle over 162,378 board feet of lumber per year, and mill about 3.4 logs or about 628 board feet a day. Annual gross revenue could reach $610,465.[5]

Orange County is one of the smaller counties in California and is the smallest of all the low-rain coastal-influenced counties.

THE HURDLES OF URBAN TIMBER PRODUCTION

Urban timber is sensible and sustainable, but there are reasons why it is not currently being practiced. Highlighted below are the five biggest hurdles to overcome for a successful urban timber program. Overcoming these barriers is the emphasis of the Design and Maintenance sections further below.

- **Capital Equipment**: Producing lumber that sells well requires a substantial outlay. Large trucks with cranes, space for storage, and buildings for milling, drying and selling are required.
- **Pricey Wood**: Urban lumber is more expensive than industrial lumber because urban areas lack economies of scale. The price of wood can be lowered through Intentional Harvest and the stabilization of quality and quantity.
- **Horticultural Diversity**: As the demand for urban timber rises, horticultural diversity might fall because marginal trees would be replaced with fast growing, high value trees, which are limited in number due to the constraints of a local climate.
- **Ongoing Harvest**: Harvesting and replanting is ongoing in a sustainable timber program. People cannot get too attached to individual trees. The yearly harvest will upset people not used to seeing continual change.
- **Public Land for Goods**: Public land is owned by all of us and ideally maintained for the enjoyment of all. Producing a product from public land that can only be bought will exclude people with limited means, meaning that not everyone gains the same benefit from the use of public resources. Urban harvest will seem exploitative to some people.

LANDSCAPE DESIGN GUIDE

Designing urban landscapes for timber production is much like designing any wooded landscape. Below are the design characteristics that ensure healthy trees and viable timber.

ACCESSIBILITY

There are many hurdles to urban timber production. One of the largest is accessibility. Getting to the tree and being able to drop it into 16-foot lengths is essential. This means

Figure 9.5   **Growing valuable timber trees requires a landscape with specific attributes. Some of these attributes are illustrated above and include flat ground, good accessibility, proper spacing, and the right type of ground cover under the trees.**

that a crane and large truck must be able to park within 20 feet of the tree. Places with great accessibility include large parks, parking lots, and streets. Commercial and residential properties generally lack easy access.

TERRAIN

The best timber comes from the straightest trunks and the straightest trunks come from flat landscapes. Planting on slopes will create wavy trunks, more so if the soil creeps and moves. If planting on slopes, then place the trees close together. The competitive nature of trees makes them stretch up for the sun and this nature has a straightening effect.

SOILS

Whether clay or sand, the ideal soil for most trees has plenty of organics, neutral to slightly acidic pH, and enough permeability to allow for the movement of gasses, nutrients, and water. Ideally, the organic content of the soil is around 3% to 6%. Most timber trees grow better with regular nutrients and water.

IRRIGATION

Not all the trees in urban environments need water once established, but many of the best timber trees do. If irrigation is required, a partial drying between watering is essential.

The wet/dry cycle passively helps exchange gases, moves salts, and ensures oxygen around the roots. Conditions that are either too wet or too dry will cause splitting, deformity, and scars.

PLANT MATERIAL UNDER TREES

The ground surface must support a tree's growth and health. An ideal ground cover will be highly permeable, low-nutrient consuming, and tough. There are two types of ground covers: Living and non-living. The characteristics of good living ground covers are described below, non-living are described further on.

Living Ground Covers for Timber Trees

The plants living under timber trees have certain characteristics that support the health of the tree: Nitrogen fixing, tough, and water conserving.

Nitrogen: In nutrient poor soils planting nitrogen fixing plants is advised. Nitrogen is the nutrient most needed by plants, and even more so for fast growing trees. In the absence of available nitrogen some plants strike up a relationship with bacteria that grab nitrogen from the atmosphere. Nitrogen fixers are mostly from the pea family and include Ceanothus, clover, lupine, sweet pea, trefoil vetch, and wallflower. Avoid nitrogen-needy ground covers, such as riparian grasses, turf, and tropical ground covers.

Tough: The plants under trees must be able to tolerate occasional foot traffic. Whether they are in public parks or parkways, people and/or their pets will be walking around trees. Some examples of good ground covers include are African daisy, cranesbill, clover, fescue, fleabane, ivy, strawberry, verbena, yarrow, rosemary, St. John's Wort, and savanna grasses, and mowed grasses for the most water needy timber trees, such as alder, ash and elm.

Water Conserving: At the height of summer, a large mature tree will transpire hundreds of gallons of water a day. Soils around large trees tend to dry fast. The plants growing under trees must be able to survive on a wet/dry watering cycle.

Non-Living Ground Covers Under Trees

The non-living ground cover must have good porosity. Some of the ideal materials include arborist chippings/mulch, angular gravel (not pea), and river rock. Decomposed granite (dg), a coarse sand-like material, is not recommended because it compacts readily and is quick to produce runoff.

PLANTING DISTANCE

Spacing can affect the quality of the timber. Too much room and a tree may twist and lean. Not enough room and trees get spindly. According to the Urban Forest Ecosystem Institute, the ideal distance for timber trees is between 70% to 90% of mature width. For example, an oak with a projected width of 30' is planted 25' to 27' feet away from another oak. Orchard farmers plan on 20' spacing, or 105 trees an acre.

However, maintaining minimum distances between trees is not always companionable with urban demands. In fact, it is rarely possible. A timberland might have 105 trees an acre, whereas a wooded, natural recreational areas may have up to 50 trees per acre; a grassy regional park may average 11 trees per acre; and single-family residential lots will have two to seven trees per acre. A sustainable urban timber program requires about 7,000 managed trees to be viable. Using this number, cities or counties need approximately 140 acres of naturally wooded areas; 636 acres of grassy regional parks; and/or 1,000 acres of committed residential lots.

LIST OF TIMBER TREES[6]

Below is a list of timber trees good for urban areas. However, growing trees with high value isn't the trick—growing timber that sells is. Always consult with local artists, furniture makers, woodworkers and any community member willing to buy wood before selecting trees. Buyers drive success.

Table 9.1:   Timber Trees

Common Name	Botanical Name	Best Uses	Harwood or Softwood/ Deciduous or Evergreen/ Height and Width/ Growth Rate
Fir, Douglas, Grand, Nordmann, Red, White	*Abies* spp.	Strong utility wood, framing, indoor and outdoor uses	Soft / Evergreen / 40–150 × 20–50/ Medium
Wattle, silver	*Acacia dealbata*	Cabinets, furniture	Hard / Evergreen / 50 × 50 / Fast
Wattle, green	*Acacia decurrens*	Landscape uses	Hard / Evergreen / 50 × 50 / Fast
Acacia, Black	*Acacia melanoxylon*	Cabinets, musical instruments, interior boats	Hard / Evergreen / 40 × 20 / Fast
Maple, big leaf and sugar	*Acer macrophyllum, A. saccharum*	Indoor aesthetics, cabinets, flooring, furniture, veneer	Hard / Deciduous / 70+ × 35+ / Medium
Alder, Black	*Alnus glutinosa*	Cabinets, furniture	Hard / Deciduous / 70 × 30 / Medium
Alder, Red	*Alnus rubra*	Cabinets, furniture, outdoor utility, utensils	Hard / Deciduous / 80 × 30 / Medium
Monkey puzzle tree, bunya-bunya, Norfolk Island pine	*Araucaria araucana, A. bidwillii, A. heterophylla*	Furniture, plywood, pulpwood, small specialty items	Soft / Evergreen / 70+ × 40+ / Medium

(Continued)

Table 9.1 (Continued)

Common Name	Botanical Name	Best Uses	Harwood or Softwood/ Deciduous or Evergreen/ Height and Width/ Growth Rate
Madrone	*Arbutus menzieii*	Tabletops, cabinets, furniture	Hard / Evergreen / 50 × 30 / Slow
Bamboo	Many genera	Indoor floors, kitchen utensils, outdoor uses, like fencing	Soft / Evergreen / 30 × 60 / Fast
Birch, yellow and red	*Betula alleghaniensis, B. nigra*	Cabinets, furniture, floors, toothpicks	Hard / Deciduous / 70 × 30+ / Fast
Cedar, Incense	*Calocedrus decurrens*	Outdoor furniture and features, decks	Soft / Evergreen / 75 × 20 / Medium
Hickory	*Carya* spp.	A utility hardwood for cabinets, furniture, sporting goods	Hard / Deciduous / 60+x40+ / Medium
Catalpa, Southern, Western	*Catalpa* spp.	Furniture, interior trim, cabinetry, carving and boatbuilding	Soft / Deciduous / 65 × 40 / Fast
Cedar, atlas and deodar	*Cedrus atlantica, C. deodara*	Outdoor furniture and structures, decks	Soft / Evergreen / 100 × 40 / Slow to Medium
Hackberry	*Celtis* spp.	Utility wood indoor and out, good for bending	Soft / Deciduous / 60 × 40 / Medium
Camphor	*Cinnamomum camphora*	Indoor cabinets, furniture, small objects, veneer	Hard / Evergreen / 60+ × 50+ / Slow
Cypress, Arizona	*Cupressus arizonica*	Outdoor furniture, decks	Soft / Evergreen / 40 × 20 / Slow to Medium
Cypress, Leylandii	xCupressocyparis leylandii	Outdoor furniture	Soft / Evergreen / 50 × 25 / Medium to Fast
Persimmon	*Diospyros kaki*	Specialty items, golf club heads, indoor trim	Hard / Deciduous / 30 × 30 / Medium
Beech, European	*Fagus sylvatica*	Indoor cabinetry, floors, utensils	Hard / Deciduous / 80+ × 50 / Medium
Ash, Arizona, white, Shamel	*Fraxinus velutina, F. americana. F. uhdei*	Cabinets, furniture	Hard / Deciduous / 80 × 50 / Fast to Medium
Cypress, Monterey	*Hesperocyparis macrocarpa*	Outdoor furniture and many uses	Soft / Evergreen / 50+ × 40 / Medium to Fast
Butternut	*Juglans cinerea*	Indoor and outdoor furniture, boxes, trim, veneer	Soft / Deciduous / 65 × 40 / Medium

(Continued)

Table 9.1   (Continued)

Common Name	Botanical Name	Best Uses	Harwood or Softwood/ Deciduous or Evergreen/ Height and Width/ Growth Rate
Walnut, black and English	*Juglans nigra, J. regia*	Many indoor and outdoor uses, coffins, furniture, flooring, sporting goods, trim, veneer	Hard / Deciduous / 60 × 40 / Medium
Juniper, western and eastern	*Juniperus californica, J. virginiana*	Indoor cabinet, furniture	Soft / Evergreen / 40 × 30 / Slow
Larch, European, Japanese and tamarack	*Larix* spp.	Many indoor and outdoor uses, construction, boxes, poles, posts, pulp	Soft / Deciduous / 60 × 30 / Medium
Larch, Western	*Larix occidentalis*	Utility wood for indoor and out, good for wet/dry conditions	Hard / Deciduous / 50+ × 15+ / Medium
Sweetgum, American	*Liquidambar styraciflua*	Indoor boxes, cabinets, trim	Hard, Deciduous / 60 × 25 / Fast
Magnolia, southern and sweet bay	*Magnolia grandiflora. M virginiana*	Indoor, outdoor cabinets, furniture	Hard / Evergreen / 50+ × 30+ / Medium to Fast
Empress Tree	*Paulownia tomentosa*	Ornamental items, furniture, musical instruments	Soft / Deciduous / 80 × 50 / Fast
Spruce, Norway and white	*Picea abies, P. engelmannii*	Construction indoor and out, fine objects, pallets, pulp	Soft / Evergreen / 100 × 40 / Fast
Pine, Bishop, Canary Island, Coulter, Ponderosa, Monterey, southern yellow	*Pinus* spp.	Anywhere indoors, plywood, pulp	Soft / Evergreen / 60+ × 35+ / Slow to Fast
Pine, eastern white	*Pinus strobus*	Masts on boats, indoor and outdoor furniture, flooring	Soft / Evergreen / 80+ × 35 / Medium
Cherry	*Prunus* spp.	Indoor furniture, cabinets	Hard / Deciduous / 30+ × 15+ / Medium
Sycamore, London	*Platanus x acerifolia*	Indoor boxes, flooring, furniture, veneer, sculped	Hard / Deciduous / 60 × 30 / Fast

(Continued)

Table 9.1   (Continued)

Common Name	Botanical Name	Best Uses	Harwood or Softwood/ Deciduous or Evergreen/ Height and Width/ Growth Rate
Spruce, Bigcone	*Pseudotsuga macrocarpa*	Utility wood both indoors and out	Soft / Evergreen / 60 × 30 / Slow to Medium
Oak, pine, red, southern live	*Quercus* spp.	Indoor use, floors, furniture, trim, veneer	Hard / Deciduous and Evergreen / 65+ × 40+ / Fast
Oak, white	*Quercus alba*	Barrels, boats, interior aesthetics, outdoor furniture	Hard / Deciduous / 80 × 70 / Slow
Locust, Black	*Robinia pseudoacacia*	Indoor, outdoor furniture, fenceposts	Hard / Deciduous / 70 × 30 / Medium
Redwood, Coast	*Sequoia sempervirens*	Many indoor and outdoor uses, load bearing	Soft / Evergreen / 90+ × 30+ / Medium to Fast
Cedar, Western Red	*Thuja plicata*	Outdoor features, decks, indoor cabinetry	Soft / Evergreen / 60+ × 25+ / Slow
Basswood, Linden	*Tilia* spp.	Musical instruments, model building, cravings	Soft / Deciduous / 50+ × 30+ / Medium
Tipu	*Tipuana tipu*	Indoor and outdoor uses, cabinets, chairs and tables.	Hard / Semi-Deciduous / 50 × 50 / Fast
Hemlock, Western	*Tsuga heterophylla*	Strong utility wood, framing	Soft / Evergreen / 150 × 40 / Slow
Elm, Chinese	*Ulmus parvifolia*	Cabinets, furniture, veneer	Hard / Deciduous / 50 × 60 / Fast

Gallery

Figure 9.6   Urban timber is rarely processed to be load bearing and consequently, it is often used for things such as benches, tables, and planter boxes. Pictured is a deodar cedar (*Cedrus deodara*) picnic bench and coast redwood (*Sequoia sempervirens*) planter box. Picture taken at Manuela's in Los Angeles, CA.

Figure 9.7   Made from deodar cedar and dynamic, this fence and shower lends themselves to a refreshing wash on a warm California day. Picture taken at Moonwater Farm, Compton, CA.

Figure 9.8   *Eucalyptus* **makes good flooring. The lumber is dense, durable and hard. Pictured are some of the types, which include (left to right) lemon scented, blue gum, red gun, and sugar gum.**

Figure 9.9   **Although milling, drying and refining timber for sale demands many machines, none of the steps are complicated. The complications are economic, personal and social. Pictured is a miller's saw at Angel City Lumber, Boyle Heights, CA.**

CAPITAL REQUIREMENTS

There is a reason why small lots do not produce and mill the trees on their property. The amount of machinery needed is too great. Below are the capital requirements for a sustainable timber program.

Notably, and as you will read in the Landscape Materials chapter, planks can be made with nothing more than a chainsaw and a warm and protected place to dry.

- Wholesale Lumber: Crane, 16' truck, mill and kiln.
- Retail Lumber: Crane, 16' truck, mill, kiln and retail center.
- Wholesale Finished: Crane, 16' truck, mill, kiln, wood shop and finish shop.
- Retail Finished: Crane, 16' truck, mill, kiln, wood shop, finish shop and retail center.

TIMBER: VALUE RELATED TO SIZE

How does one know if a tree has timber potential? While value is often assigned to a type of tree, typically it is the size and straightness of the trunk that ultimately determines its value. As a rule:

- Best Wood: 16-inch diameter at the small end and at least 8 feet long.
- Good Wood: 14-inch at the small end and at least 8 feet long.

Figure 9.10 Although sweetgum (*Liquidambar* spp.) produces lumber with exceptional quality, this tree is of low value because of its small size. If it can grow straight and avoid serious injury for another 5 to 10 years, it will have high value. The placement of this tree makes it relatively easy to cut down and haul away for milling.

- Fair Wood: 12-inch at the small end and at least 6 feet long.

Two of the leading resources for determining the value of urban timber are:

- *Recycling Municipal Trees: A Guide for Marketing Sawlogs from Street Tree Removals in Municipalities*. Edward T. Cesa, Edward A. Lempicki, J. Howard Knotts. Dept. of Agriculture, Forest Service, Northeastern Area State & Private Forestry, Forest Resources Management. 2003.
- *Harvesting Urban Timber*. Sam Sherrill. Linden Publishing. 2003.

Maintenance of Timber Trees

Ensuring valuable timber trees means managing tree health, maintaining straight trunks, minimizing damage to the trunk, and the constant removal and replanting of the urban forest.

REMOVAL AND REPLANT

Timber trees get harvested every 15 to 20 years. Harvesting trees at 15 years means that 1/15 of the managed forest gets removed every year. A 10,000-tree program would require the removal and replanting of 667 trees a year.

Large public parks have an advantage over parking lots, parkways and streets, because the replanted trees do not have to go in the same spot. Removing and replanting trees in parkways, for example, means removing the stump, which can increase costs by 30%. In these situations, stumps are either dug out, grinded out, or drilled out.

OBJECTS

One of the largest disadvantages with milling urban wood is the objects found inside the tree. From nails and rocks to cables and fences, trees can grow around, into, and over all types of urban artifacts. In fact, trees can completely absorb a rock or object with little or no outward appearance. Along with reducing quality, these objects will ruin saw blades and endanger the people milling.

Maintenance is aimed at removing and deterring these objects. Public education may be necessary. People in public parks will need to be asked not to screw or staple things into the trees. Pruning makes a difference, too. Trees with tight crotches, those that resemble a V, are more prone to collect objects. Tight Vs are also structurally weaker than wide angle branch unions.

Most millers of urban timber give each log a visual inspection and might pass a metal detector over it before milling. They are looking for any outward signs of objects. Black and blue stains outside or inside the wood usually indicate metal. Unfortunately, not all

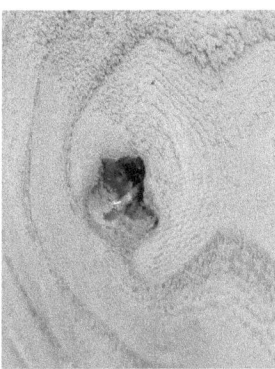

Figure 9.11 **Luckily, the sawyer's blade was strong enough to slice through this binding wire, although it was duller as a consequence. Concrete, rocks, and wire are common hazards when milling urban timber.**

urban objects are metal. Filling wounds with concrete was a best management practice in the 1970s and 1980s and is a sawyer's nightmare.

PRUNING

The goals for pruning timber trees are to promote health, maintain one leader trunk, and ensure a tree's energy is going to growing a large, tall and straight trunk. Use the rules below to ensure healthy, valuable timber trees.

Public Health: Opposed to forest-grown timber, trees in urban areas will need pruning to protect the public's health. Falling branches are a genuine concern for people living in vegetated environments. From the simple inconvenience and sometime small scare, to downed power lines and extensive property damage, trees demand maintenance to prevent such tragedies. In some cases, the likelihood of facture is visible before the actual break. Simple inspection is the key.

Never use the techniques below to remedy trees' problems. These techniques are harmful and will cause additional unwanted problems.

Limbing Up: Prematurely removing lower branches creates what is called a "kite". Limbing up slows the growth of the circumference of the trunk while accelerating its height, ultimately reducing the size and value of the timber and creating a greater degree of fracture risk in high winds.

Topping: This is the practice of cutting off the top of a tree. This practice creates an abundance of suckers and watersprouts, which increases the number of knots in the lumber. Furthermore, topping can reduce the mercantile height.

Pruning Without Reason: Many trees in urban areas are pruned at regular intervals, whether they need it or not. However, reducing a tree's canopy reduces photosynthesis, which in turn reduces biomass creation and the size of the timber. More photosynthesis means more lumber.

Pruning More Than One-Third: Removing more than 1/3 of a tree while pruning can severely stunt its growth, if not kill it. As a rule, never prune more than 20% of a tree.

Bad Timing: Typically, hard pruning is done during a tree's dormant season, which varies greatly between species. As a rule, deciduous trees are pruned in winter, Mediterranean plants in early fall, and tropical plants after flowering. A certified arborist should always be consulted.

FERTILIZE

The rule for fertilizing trees is .15 pounds of actual nitrogen for every inch in trunk diameter, as measured at breast height. For example, using a fertilizer with 10–8–7 on a tree with a 10-inch trunk requires 1.5 pounds of fertilizer. Typically, the 1.5 pounds is divided in half and one half is used in spring and the other in late summer. Fertilizers are kept 6 inches from the trunk and spread over an area 1.5 times the canopy. Additional supplements, like sulfur, may be needed as well.

PEST CONTROL

If pests affect productivity and/or the quality of wood, then swift action is required. The problems most affecting trees are leaf chewing insects, such as budworms and gypsy moths; insects that bore into the wood or between the bark and wood, like beetles; and fungi, such as blight.

Chemical insecticides are commonly used on trees because physical controls, like sticky tape, are difficult to apply to large trees. But before resorting to chemicals make sure the problem is not linked to a cultural practice. Too much water and a tree will be prone to fungus problems. Too little water and the tree will not be able to repel boring insects with sap. Too much fertilizer and trees produce nutrient rich, insect attracting growth.

If a pest problem is persistent and a tree constantly needs application of any type of control, then it is the wrong tree for that location/region and should be removed and replanted with a better adapted species.

MONITORING

An intentional urban timber program may not be an immediate success but can become one with constant monitoring and bookkeeping. Keeping records is the key to sustained and long-term success. Monitoring will help determine the fastest growing trees, the least expensive to maintain, the most pest-free species, and the trees with the least impact on urban infrastructure. Bookkeeping will also reveal the trees with the greatest value, the type of boards and cuts that sell the best, and the most cost-effective levels of production (rough lumber or finished products).

CITIES NEED TREES

People are biologically, economically, and psychologically connected to trees. They provide a wide range of incredible benefits. Urban timber production can increase the number of trees, magnifying the benefits listed below.

Indoor Energy Use: Strategically placed trees, shrubs and ground covers can reduce the energy needed to cool and heat structures by as much as 50%.[7]

Crime Reduction: The number and density of urban trees is correlated to rates of crime: more trees, less crime.[8]

Pollutant Removal/Public Health: Trees can reduce air and noise pollutants. Research shows that trees reduce the number of particulates and gaseous pollutants—both of which are carcinogens and reduce respiratory health. Trees also provide some buffer against noise and glare, and the thicker the vegetation, the greater the buffer.[9]

Personal Health: Landscapes can reduce a person's heart rate and blood pressure, increase feelings of security and wellbeing, and improve learning outcomes and recovery from injury. Simply put—trees are great for public health.[10]

Public Safety: Trees and shrubs along roads slow drivers. Trees reduce car crashes and increase biker and pedestrian safety.[11]

Thermal Comfort: Trees will reduce the temperature you feel outside by 3°F to 5°F.[12] This is a significant reduction. If trees shade surfaces that store heat energy, such as asphalt and concrete, they reduce nighttime temperatures as well.

Stormwater Management: Urban forests stop, slow and clean urban runoff. They decrease topsoil loss, remove pollutants, and protect waterbodies. Evergreen trees are far more effective than deciduous trees.[13]

Property Value: Trees and greening will improve the aesthetics of a community and consequently, have a positive and large impact on property values.[14]

Sequestration: Per square foot trees capture 45% more carbon than any other landscape type in urban environments.[15] But trees' superpower is not their sequestration of carbon, it is their ability to hold on to it for a century or more.

Conversions: Unlike other landscape types, trees can be converted into more useful things before returning to their constituent states, which are gases and minerals. As an example, a tree can be turned into lumber and when the lumber product has expired, it can be turned into landscape materials, energy, fertilizer and/or mulch. According to John Lyle, FASLA, conversion is fundamental to the success of an urban ecosystem.[16]

Resources

BOOK

Harvesting Urban Timber. Sam Sherrill. Linden Publishing. 2003.

WEBSITES

Forest Resources Association. Can be viewed at https://forestresources.org/
Society of American Foresters. Can be viewed at https://www.eforester.org/
U.S. Forest Service, *Urban and Community Forestry Program.* Can be viewed at https://www.fs.usda.gov/ managing-land/urban-forests/

Notes

1 Nowak, David J., et al. "Annual Biomass Loss and Potential Value of Urban Tree Waste in the United States." *Urban Forestry & Urban Greening*, vol. 46, December 2019. https://www.sciencedirect.com/science/article/abs/pii/S1618866719303656?via%3Dihub Accessed March 18, 2023.

2 Kent, Douglas. 2010. *Timber Production in Southern California: An Economic and Carbon Analysis.* Master's Thesis. Landscape Architecture, California Polytechnic University, Pomona p. 47.

3 McPherson, Gregory, et al. "Urban Forestry and Climate Change. Center for Urban Forestry Research." USDA Forest Service, Pacific Southwest Research Station, 2013. https://www.fs.usda.gov/ccrc/topics/urban-forests Accessed March 18, 2023.

4 Douglas Kent. 2010. *Timber Production in Southern California.*

5 Douglas Kent. 2010. *Timber Production in Southern California.*

6 Douglas Kent. 2010. *Timber Production in Southern California.*
     Sherrill. *Harvesting Urban Timber.* Fresno, CA: Linden Publishing. 2003.
     Meier, Eric. *The Wood Database*, 2023. https://www.wood-database.com/wood-filter/ Accessed March 18, 2023.
     SelecTree, Urban Forest Ecosystem Institute, California Polytechnic University, San Luis Obispo. https://selectree.calpoly.edu/ Accessed January 15, 2023.

7 Gregory E. McPherson, et al. "Impacts of Vegetation on Residential Heating and Cooling." *Energy and Buildings*, vol. 12 no. 1, April 24, 1988, pp. 41–51. https://www.sciencedirect.com/science/article/abs/pii/0378778888900540

8 Neckerman, Kathryn M., et al. "Disparities in Urban Neighborhood Conditions: Evidence from GIS Measures and Field Observation in New York City." *Journal of Public Health Policy*, vol. 30. March 2, 2009, pp. 264–285.

9 Gorsevski, Virginia, et al. 2002. "Air Pollution Prevention through Urban Heat Island Mitigation: An Update on the Urban Heat." *Island Pilot Project, Draft Report*, Lawrence Berkeley National Laboratory, Berkeley, 2002. https://www.coolrooftoolkit.org/wp-content/uploads/2012/04/epa_doc.pdf
     McPherson, Gregory. "Chicago's Urban Forest Ecosystem." U.S. Department of Agriculture, Forest Service. 1994.
     Rowntree, Rowan A. "Ecological Values of the Urban Forest." In *Proceedings of the Fourth Urban Forestry Conference*, Washington, D.C., American Forestry Association, 1989, pp. 22–25.

Nowak, David. J and Crane, Daniel E. "The Urban Forest Effects (UFORE) Model: Quantifying Urban Forest Structure and Functions." *Integrated Tools for Natural Resources Inventories in the 21st Century: Proceedings of the IUFRO Conference*, ed. M. Hansen, T. Burk. U.S. Department of Agriculture, Forest Service, North Central Research Station, 2000, pp. 714–720. https://www.fs.usda.gov/research/treesearch/18420

"Study Shows Benefits of Urban Trees." *Journal of Environmental Health*, vol. 63, no. 4, November 2000, p. 40. https://www.jstor.org/stable/i40190120

Jo, Hyun-Kil and McPherson, Gregory E. "Carbon Storage and Flux in Urban Residential Greenspace." *Journal of Environmental Management*, vol. 45, June 14, 1995, pp. 109–133.

10 Ieszic Formeller. *A Healing Home: The Application of Therapeutic Landscape Design Theory to the Residential Setting.* Master's Thesis accepted by California State Polytechnic University Pomona, Department of Landscape Architecture. June 2010.

and

Jaret, Peter. "Harvesting Benefits of Trees. "*National Wildlife*, vol. 37 no. 5. August 1999, p. 12.

11 Black, Tom. "Slow Ride: Calming Traffic." *American City & County*, vol. 113 no. 9. August 1998, p. 30.

and

Olson, Bob. "Trying to slow urban traffic." *Futurist*, vol. 24 no. 3. May 1990, p. 74.

12 Shashua-Bar, et al. "The Cooling Efficiency of Urban Landscape Strategies in a Hot Dry Climate." *Desert Architecture and Urban Planning*, The Jacob Blaustein Institutes for Desert Research, Ben-Gurion University of Negev, Israel, vol. 92 no. 3–4, September 30, 2009, pp. 179–186. http://dx.doi.org/10.1016/j.landurbplan.2009.04.005

13 Kent, Douglas. *Ocean Friendly Gardens. A How-To Gardening Guide to Help Restore a Healthy Ocean and Coast.* Surfrider Foundation 2009.

14 Dwyer, John F. "Economic Benefits and Costs of Urban Forest." *Alliances for Community Trees: Proceedings of the Fifth National Urban Forest Conference.* American Forestry Association, Washington, D.C., 1991, pp. 8–58.

15 Bassham, James A. 1980. "Energy Crops (Energy Farming)." *In Biochemical and Photosynthetic Aspects of Energy Production*, ed. Pietro, Anthony San, Academic Press, pp. 249–263. New York, NY.

and

Kirkham, Tim. *Sustainability in Landscaping.* City of Irvine, CA, 1989, Appendix p. 4.

16 Tillman, Lyle, John. *Regenerative Design for Sustainable Development.* John Wiley & Sons, Inc. 1994.

Chapter 10

Water: Greywater Capture and Use

The use of greywater is headed in one of two directions. It will either be an essential part of our sustainable future, or it's going to be just another excuse to sell plastic contraptions. It either produces more benefit than cost, or creates more air, water and waste problems. The trick with greywater is to ensure that benefits exceed the costs of capturing and distributing the water.

Covered below is an overview of greywater systems. This chapter covers the ideal types of properties for greywater reuse, the characteristics of an effective system, lists some of the many plants that thrive with greywater, and highlights the maintenance needed to operate a system whose benefits exceed its costs.

Greywater Overview

Uses: The daily discharge is used to grow various landscape products and services.

Costs: Low to high, depending on the type of system.

Difficulty: Not difficult to moderately difficult; expertise needed includes plumbing and general building.

Best Scale: The best site for a greywater system is a property that genuinely needs more irrigation water. The site may have water restrictions, or not be connected to a municipal supply, or be on a limited supply of ground water. Simple systems are preferred on low discharge properties (hundreds of gallons a day), such as single-family residences. Complex systems are better suited for large discharge properties (thousands of gallons a day), such as manufacturing and hotels.

ERoEI: Energy gains from greywater can easily get buried under a pile of plastics, metals, and good intentions. Many of the materials in greywater systems produce air and water

DOI: 10.4324/9781003369752-10

pollutants during manufacturing (PVC), and represent a load of embedded energy and carbon. Plumbing-focused systems can push environmental costs well beyond the reach of any benefit. The goal is to create an energy benefit, such as cooling or food, which exceeds the energy cost of the system.

Notes: Since greywater can't be stored and the landscape receives a daily dose, the challenge is ensuring a benefit from this nutrient rich deluge, not just dumping the effluent in the landscape. Greywater should be used to achieve an ecological benefit, be that more community cooling and thermal comfort, more food or products, or more biodiversity and public health. This is no easy task.

Nomenclature

Blackwater: Discharge water that has a high chance of having pathogens, metals and/or toxins. Blackwater is illegal to capture onsite. Common sources of blackwater include toilets and industrial effluent.

Complex System: A large and permitted greywater system that typically has tanks, filters, pumps, and irrigation devices.

Gravity Fed: The distribution of greywater using gravity.

Greywater: There isn't one definition of greywater—there are many. Some organizations claim that greywater is "any wastewater not generated from toilet flushing." Others are more restrictive and exclude kitchen water (sink and dish washer). No matter the definition, greywater does not contain a great amount of either pathogens or toxins.

Non-Potable: Water that is unfit to drink, such as reclaimed water.

pH: A figure that represents the acidity or alkalinity of something, such as soil and water. Greywater tends to be alkaline.

Plumbing Code: Whether state, county or city, nearly every jurisdiction within the U.S. has a plumbing code that dictates the minimum requirements for constructing and maintaining a greywater system.

Potable: Water that is fit to drink.

Receiving Landscape: The garden area where greywater is discharged.

Simple System: A low-tech greywater system commonly found on residential properties. Most are gravity fed.

Soap: Whether shower, sink or clothes water, the soaps mixed with greywater affect the health of the receiving landscape. As a rule, avoid soaps with boron, chlorine, petroleum distillates, and high salts.

Soil Type: The percentages of clay, silt and sand in soil. Soil type determines how fast a soil can absorb water, hold on to nutrients, exchange gasses, and facilitate a healthy living system.

Best and Worst Types of Properties for Greywater

Best Properties

There are some properties where reusing greywater makes economic and ecological sense.

Knowledgeable, Committed People: A greywater system is only as good as the knowledge of the people that benefit from it. If they don't understand the objective the system is aiming to achieve, they won't be motivated to maintain it. Without educated and committed people, the quality of effluent declines, the plumbing gets clogged, and the landscape struggles to produce a measurable benefit.

The same thing applies to the landscape the greywater system was designed to serve. It also needs to be well cared for.

Strong Need: A majority of urban areas are plumbed to a municipal supply of potable water. Managing greywater, a non-potable water supply, makes the most sense on properties where there is great need for more water; where greywater can pay for the infrastructure needed to manage it. Some of the highest dividends are:

- *Community Cooling*: A large-broad leafed tree can cool a hot summer day by as much as 6°F, although a lot of water is needed for the feat. At the height of summer, a large tree may transpire more than 300 gallons of water a day. Greywater is perfect for growing large trees that cool communities.
- *Production Landscapes*: Whether veggies or fruit trees, fibers or timber crops, production orientated landscapes will generally benefit from the added water and nutrients. What little public health risk greywater poses can easily be mitigated by common sense practices.
- *Fire Country*: Living in wildfire country is a challenge. Water is scarce and water-loving plants are discouraged, despite their impeccable record for repelling firebrands and wildfires. Greywater can increase the number of water-hungry, fire-retarding plants.
- *Wetland Habitat*: Vernal ponds, mudflat flora, and fresh water tidal habitat can benefit from the use of greywater. The increase in wet areas can help encourage a greater range of plants and animals.

Worst Properties

Greywater can get dreadful fast, and there are properties and people that should never install a system.

Unsuitable Landscape: Greywater is ill advised for properties that are too steep (it increases soil instability); too cold (it freezes or stagnates); too wet (it stagnates); too dense of soil (it stagnates); and/or too small to handle the discharge (it overflows offsite).

Inadequate Maintenance: The maintenance mantra for greywater, like so much of the other work in this book, is a little at a time, all the time. Greywater systems need a strong commitment to maintenance—lack of it can lead to poor ERoEIs and public health problems.

Costs: Even the simplest greywater systems have costs. Pulling effluent away from a sewer or septic system and pushing it to a landscape involves a variety of energies and materials: Electricity, gasoline, concrete, metals, plastics, and PVC are common. A greywater system should not be installed on an urban lot if costs exceed benefits.

Types of Systems

In both legal and practical terms there are four types of greywater systems: Clothes washer, simple, complex and treated.

Figure 10.1   **Not every property has the right characteristics to make the most of greywater. A good property will have ample space, porous soil, productive uses for the water, and diligent people. A bad property will have just the opposite.**

Figure 10.2 **Three of the most common greywater systems in urban areas are laundry to landscape, simple and gravity fed, and complex and pumped.**

Clothes Washer: Most of the nation's greywater systems use washing machine water. Little wonder—the water is pumped, can travel up small inclines, and the system requires little plumbing—all of which makes irrigating a landscape easy.

Simple: If the daily discharge is 250 gallons or less, then the greywater system is considered Simple. These low-tech systems are mostly unpermitted and residential. Simple systems abscond the bells and whistles, are gravity driven, and are plumbed from a variety of greywater sources, such as washing machines and bathroom sinks. The irrigation devices are straightforward too and include flooding, perforated drainpipe, and garden hoses.

Complex: Designed for safety and flexibility, complex systems share many of the same components as rainwater harvesting and septic systems. They too include tanks, filters, pumps, and extensive plumbing. These large and permitted systems are designed for year-around discharge. Because of the filters, greywater can be pumped through multiple irrigation valves and a variety of low-flow irrigation devices, such as drip emitters. LEED certification typically demands complex systems.

Treated Greywater: Using greywater indoors or for overhead irrigation is illegal. The possibility of spreading pathogens had been deemed too great. However, if greywater is run through a sterilization process, then it can be used indoors. This water is called Reclaimed or Purple Pipe Water.

System Overview

The essential attributes of a system's plumbing and receiving landscape are covered below.

Figure 10.3   **Wastewater that is highly treated and reused is called reclaimed water. Pictured is the reclamation plant at Stone Brewing in Escondido, CA. They are one of the most water efficient breweries in the nation because they treat and reuse most of their greywater.**

Plumbing[1]

Provided below is an overview of capturing, moving, and irrigating with greywater.

CAPTURING GREYWATER

Just a hundred years ago buckets would have been used to capture greywater. There would have been pails in the laundry room, kitchen and bathrooms. Despite its efficiency and low cost, buckets are hardly convenient, and they might even be illegal. Modern systems are ruled by plumbing.

Laundry water is the easiest to capture and typically plentiful. Bath and shower water are usually harder to intercept because the plumbing is at ground level and it may be too close to toilets. And while the plumbing for most faucets is accessible, hooking all the faucets together can be a challenge. As a rule, water coming from elevated areas, like second or third stories, is captured first because of its gravitational and embodied energy.

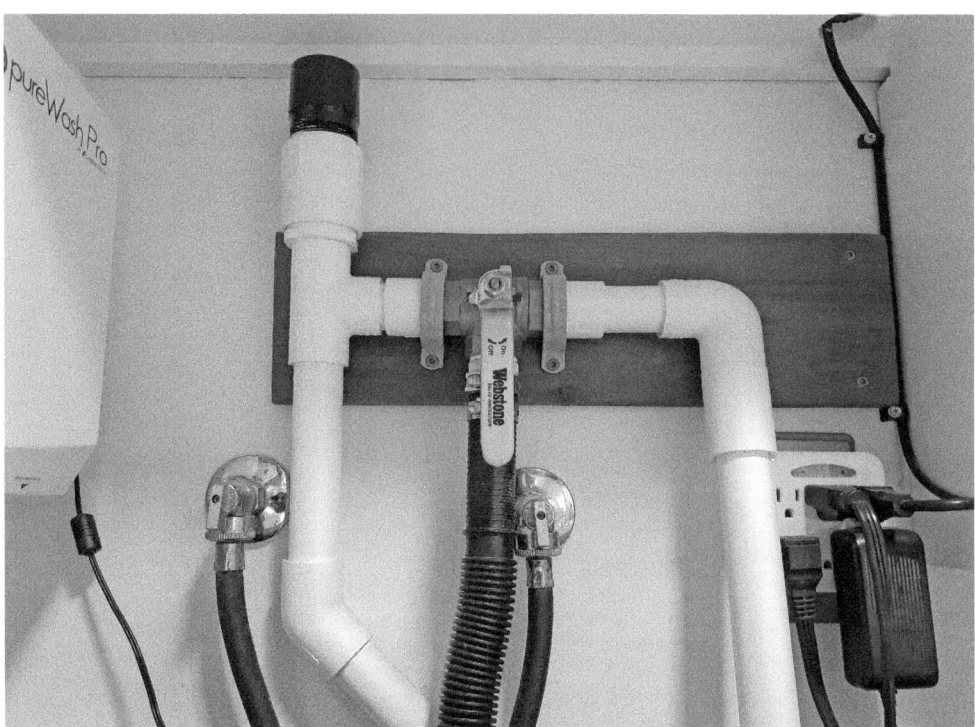

Figure 10.4 **Pictured is a simple laundry to landscape system. The pipe in the middle discharges the washer's water. The pipe to the left leads to the landscape and the one to the right the sewer. The yellow valve lets the owner switch between the two, depending on irrigation needs. The black device in the upper left allows airflow, releases gasses, and helps prevent the water from backflowing.**

Sources and Volumes of Greywater

The average U.S. household of 2.6 persons generates 101 gallons of reusable water a day, 38.7 gallons per person. Below is a chart showing common sources of greywater and their approximate volumes.[2]

Table 10.1: Sources and Volumes of Greywater in the US

Source	Amount in Gals.	Percent of All Indoor
Blackwater: Dish washer, kitchen sink and toilets	30.6	44.2
Greywater: Bath, clothes washer, faucets, shower	38.7	55.8
Breakdown		
Bath	1.2	1.7
Clothes Washer	15	21.6
Dishwasher	1	1.4
Faucets	10.9	15.7
Leaks	9.5	13.7
Other Indoors	1.6	2.3
Shower	11.6	16.7
Toilets	18.5	26.7

Many states calculate the volume of greywater based on formulas. For example, the state of California uses the figures above to estimate volumes of greywater: Showers, bathtubs and wash basins produce 25 gallons per day per occupant; laundry machines produce 15 gallons per day per occupant:[3]

MOVING GREYWATER

The plumbing required to circumvent the sewer mirror those of any effluent discharge: The pipes are no less than 1.5" ABS (2" preferred); Y-valves are installed so that every source can be switched from the landscape to sewer; the connections, corners and bends are wide and sweeping; cleanouts are provided in easy-to-access places; gas vents are installed along the discharge pipe; and there is absolutely no link to a potable supply of water.[4]

COLLECTING GREYWATER

Greywater cannot be stored. It gets too foul, too fast. Storing water for over 48 hours is taboo; no more than 24 is recommended. However, greywater can be collected to make pushing it to a landscape easier. These collection barrels are called Surge Tanks. A surge tank can move water further from a structure than a gravity fed system; accumulation creates more pressure and opportunity.

Typically, the average household needs nothing more than a 55-gallon drum. Size is based on the highest expected inflow of greywater. For instance, if a washing machine, shower and faucet were all discharged at the same time, then the inflow would be about 31 gallons of greywater. Storage tanks are designed to either passively or actively discharge when full. Simple systems use a closing siphon and complex systems use a pump with a float.

IRRIGATING WITH GREYWATER

Irrigation can be the trickiest part of a greywater system. Below are the basics of simple and complex systems. No matter the type, there are two rules that apply to every system:

1. Keep it simple.
2. All greywater irrigation is done subsurface.

Simple Irrigation

Low tech is gravity driven, going straight from the structure to the landscape. Installation and maintenance costs are low. But there are drawbacks. Unless on a slope, greywater is only as transportable as a 2% slope, which usually puts the greywater close to its source: The structure. There are three primary ways to irrigate with gravity-driven water: Perforated drainpipe, mulched basin (flooding), or French drain (trench).

OVERVIEW OF A SIMPLE SYSTEM

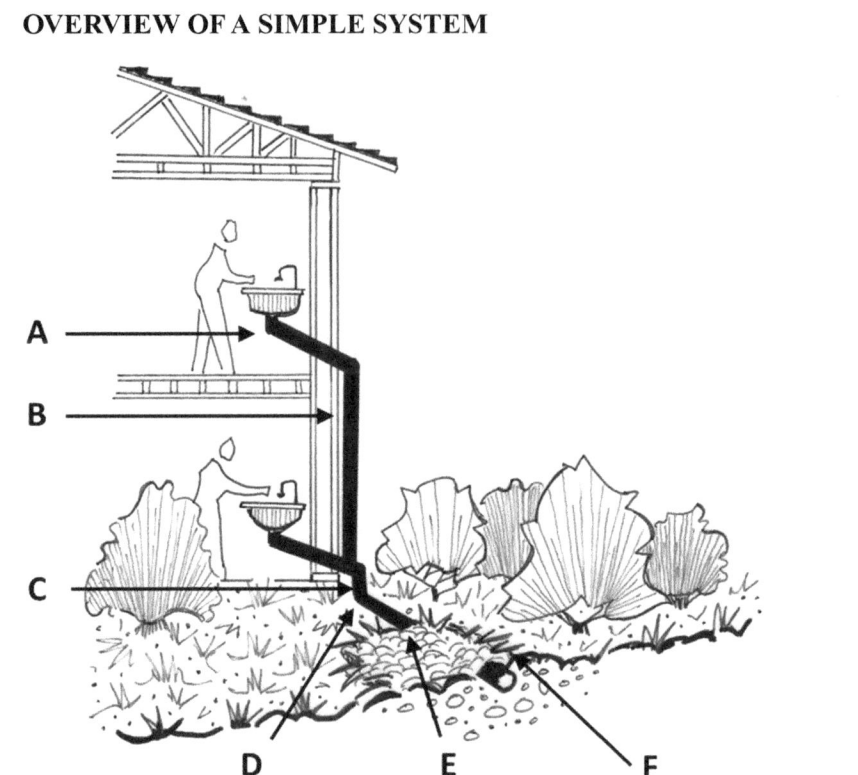

Figure 10.5   **A: Entire system is held together through compression, not glue. B. Passively lead water is led in drainpipe no less than 3". The pipe should be high-density polyethylene, ABS, or PVC. C. All corners and turns are made with wide 45° and 90° sweeps. D. No point of the entire system has less than a 2% decline. E. Discharge area is 10 feet away or downhill of structure. F. Mulch is no less than 2" to 4" thick, shielding greywater from human contact.**

Complex Irrigation

Large-tech irrigation systems have many advantages over low tech. Backed by a surge tank and pump, the water is more transportable. These systems may run multiple irrigation valves. And, because the water can be delivered at higher pressures, they can use an array of irrigation devices. Emitter tubing, soaker hose, drip tubing, and conventional flooding are the primary methods of irrigation. Naturally, the largest drawbacks with large tech are the environmental and economic costs of installation, operation, and maintenance.

Landscaping with Greywater

Designing a productive greywater landscape entails managing three variables: the property, the size of the receiving landscape, and the plants. The basics of each are listed below.

IDEAL SITE CHARACTERISTICS

The ideal landscape for receiving greywater will have six attributes.

- **Access**: The discharge area must have ease of access. Healthy thriving landscapes require regular cleaning, weeding, and renewing.
- **20 Feet of Distance**: The receiving landscape should be 20 feet and/or downhill from a structure. The goal is to protect the structure from moisture.
- **2% Grade**: Gravity powered systems require a minimum 2% fall to move the water; 3% to 4% is much better.
- **Size of Discharge Landscape**: The discharge area must be sized for the estimated maximum discharge volume. Keep reading.
- **Plants**: Greywater landscapes are often densely planted. More plants mean more photosynthesis and transpiration, which in turn reduces chances of soil saturation and the accumulation of chemicals, metals and salts.
- **Emergency Exit**: It is not unusual for greywater landscapes to flood during heavy rain. An emergency exit for excess water must be a part of the system. These exits can either be drainpipes or trenches. Exits typically lead to storm drains, another landscape, or sewer.

SIZING THE RECEIVING LANDSCAPE

In theory sizing a landscape for greywater discharge is simply a numeric prescription based on soil type and amount of effluent. In practice, it is much more complicated. Exposure,

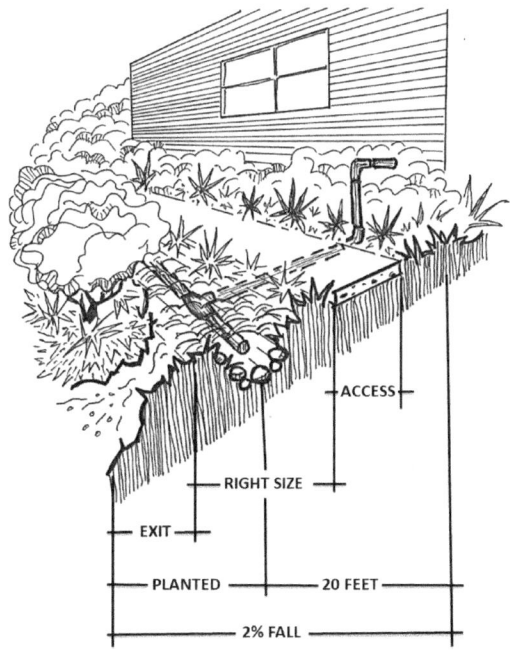

Figure 10.6   **Overview of a good landscape**

Table 10.2:   Sizing the Receiving Landscape for Greywater

Type of Soil	Gls. per S.F. per day[1]	Square feet needed per 100 Gls[2]	Square feet needed per 1 Gl.[3]
Coarse sand or gravel	5.0	20	0.2
Fine sand	4.0	25	0.25
Sandy Loam	2.5	40	0.4
Sandy clay	1.7	60	0.6
Clay with considerable sand or gravel	1.1	90	0.9
Clay with little sand or gravel	0.8	120	1.2

1. Gallons: Maximum absorption capacity in gallons per square foot for a 24-hour period.
2. Sq. Ft: Minimum square feet per 100 gallons of graywater discharged per day.
3. Minimum square feet of area per 1 gallon of graywater discharge per day.

humidity, plant types, season, organic accumulation, temperature, and wind all influence a landscape's ability to make use of greywater.

The figures on sizing below come from the State of California's Plumbing Code.[5] California's figures error on the side of caution. A mulched and densely planted sandy depression can handle more than 5 gallons of greywater per day per square foot. Start with the figures below, but over time a better understanding will evolve.

SOME OF THE BEST PLANTS FOR GREYWATER SYSTEMS

Ornamental plants are not included in the lists below; these lists strive for more tangible outcomes, such as urban cooling, food, products, timber, and biological diversity. As a rule: Acid loving plants struggle with greywater; deciduous plants do better than evergreen; and plants found naturally along dry streams that seasonally flood will thrive.

The plant lists were developed from multiple sources.[6] Studies on the effects of greywater have found that most plants are not greatly affected, especially if greywater is supplemented with fresh. If greywater affects a plant, then it was probably a result of salt. The following lists are as much about salt tolerance as tolerance of greywater.

Community Cooling

The amount of cooling is related to the degree of shading. The degree of shading is related to density and size of the tree. The density/size of tree is related to amount of water needed. Using greywater to cool structures and communities instead of imported highly treated potable water can be an excellent, energy-wise strategy.

Ash, most (*Fraxinus* spp.)
Camphor (*Cinnamomum camphora*)
Carob (*Ceratonia siliqua*)
Chinese Elm (*Ulmus parvifolia*)
Chinese Pistache (*Pistacia chinensis*)
Ficus (*Ficus* spp.)

Oak, larger and deciduous varieties (*Quercus* spp.)
Shower Tree, Medallion (*Cassia* spp.)

Food

A lot of crops thrive with abundant water and nutrients, but not all are compatible with greywater. Fresh leaf and root crops, like lettuce and carrots, are not recommended because of the possible contact with pathogens. Instead, the harvestable edible should be at some end of a vascular system, like flowers, fruits, and nuts.

Annuals: Artichoke, cucumbers, peppers, tomatoes, zucchini.

Berries: Blackberry, blueberry, raspberry, strawberry.

Deciduous Fruit: Almond, fig, grape, plum, persimmon, walnut.

Herbs: Comfrey, horehound, lemon balm, lavender, mint, rosemary, sage, *Viola*.

Tropical: Banana, cherimoya, date palm, guava, mango, passion fruit.
    Crops sensitive to greywater (salts) may have to be avoided, such as avocados, citrus, various herbs, and seedlings.

Figure 10.7   Passionflower (*Passiflora edulis*) grows like a weed with almost any type of water, greywater included. Both the fruit and leaves are edible. Pictured with this flower is a green fruit eater beetle, which enjoys fallen fruit.

Product

Greywater landscapes can be a large part of the craft movement in the U.S. It can be employed to grow dyes, medicines and timber.

Holiday: Holly, juniper, mesquite, palm.

Dye: Calendula, cherry, Chinese tallow (*Triadica sebifera*), elderberry, oak, redbud, tree of heaven.

Fiber: Bamboo, cotton, henna, palm, willow.

Medicinal: *Aloe vera*, *Senna*, dandelion, elderberry, *Echinacea*, *Gingko*, mint, St. John's Wort, *Vitex*, yerba mansa.

Oil, Resin, Sap: Birch, camphor, castor bean, *Eucalyptus*, maple, *Melaleuca*, palm, tea tree.

Timber

Timber trees are an ideal crop for greywater. Not only do the trees provide valuable lumber, but they can also sustain birds, cool communities, and protect waterbodies—all of which helps a landscape reach carbon neutrality.

> Acacia, black (*Acacia melanoxylon*)
> Birch, red (*Betula nigra*)
> Elm, Chinese (*Ulnus parvifolia*)
> Maple, big leaf (*Acer macrophyllum*)
> Locust, black (*Robinia pseudoacacia*)
> Oak (*Quercus* spp.)
> Sweetgum (*Liquidambar* spp.)

Wetland, Constructed

Wetlands can provide two big benefits in urban areas: cleaner water and biological diversity. Constructed wetlands are particularly effective at cleaning greywater of chemicals, metals, nutrients, and pathogens.[7] And if managed properly, a wetland can be a windfall for the insects and birds that thrive within flooded systems—ecosystems that are sorely lacking in urban areas.

> Alder (*Alnus* spp.)
> Cattail (*Typha* spp.)
> Aspen, Cottonwood and Poplar (*Populus* spp.)
> Iris, bog type
> Rush (*Juncus* spp.)
> Plane Tree and Sycamore (*Platanus* spp.)
> Willow (*Salix* spp.)

Notably, there are many edible species that can be grown in constructed wetlands, including bulrush, cattail, celery, crawfish, miner's lettuce, snail, speedwell, and watercress.

Maintenance Practices

The degree of ecological benefit is enhanced through regular maintenance. Interestingly, greywater is one of the few land management practices that demands as much attention inside a structure, ensuring the highest quality water, as it does outside a structure, ensuring that all the water goes to its greatest use. Below are the essentials of greywater maintenance.

Plumbing

Filters: Filters should be maintained to ensure a minimum flow of 25 gallons per minute. If greywater has a lot of solids, then filters will need to be cleaned once a month.

Irrigation: Large discharge pipes become havens for insects, and irrigation systems become clogged. The remedy for both is a vigorous flushing with potable water once a year.

Mulch: Greywater landscapes are mulched for a variety of reasons. The most important is to protect people from accidental contact with the water. Organic mulches also reduce the accumulation of salts and stabilize a soil's pH. Organic mulches may need replenishing as often as every 4 months, but at least once a year.

Storage/Surge Tanks: Storage tanks will need cleaning twice a year. Algae and the accumulation of solids will build up and reduce a system's ability to move water. Cleaning tanks also reduces odors.

Discharge Depressions

Whether through erosion, the accumulation of organic matter, or the laws of gravity, low spots in a landscape fill. Maintaining the depth of the receiving landscape maintains the system's capacity. As a rule, depressions should be cleaned and dug out when they lose 10% of their capacity, which is approximately every 2 to 4 years.

Soils

Greywater affects soil and not all the effects are good. The largest problems with greywater are salt accumulation, alkalinity, nutrient loading, and soil compaction.

HANDLING THE BUILDUP OF SALTS

Soil collects salts. They accumulate in arid regions where evaporation exceeds precipitation; in gardens victimized by over-fertilization; in landscapes where the irrigation is shallow and frequent; and in landscapes irrigated by reclaimed or greywater. The impacts of salt accumulation are diverse and include desiccation, chlorosis, leaf burn, leaf drop, withering, yellowing, and poor quality and quantity of flower, fruit, and seed. Here are some remedies:

- **Leach**: Drench an area with fresh water to push the salts beyond roots. Leaching demands a lot of water. No less than 6 inches of water is needed to properly leach an area, which is about 750 gallons for 200 square feet. If annual rainfall is less than 18 inches a year, leaching may be a necessity.
- **Humus**: Regularly apply humus/finely composted mulch over discharge area. Humus is great for soil health. It is acidic and helps correct the alkalinity of greywater. It inspires microscopic life and increases porosity. And it is rich with plant available nutrients.
- **Replant**: Replace salt-intolerant plants with those that thrive with greywater and its salts. See the plant lists above.

ALKALINITY AND IRON DEFICIENCIES

Greywater is typically alkaline. This sweet (alkaline) water creates sweet soils which lock up iron, making it unavailable to plants. An iron deficiency shows itself as chlorosis. Remedies include:

- **Humus**: Applying rich humus twice a year will slowly acidify the soil
- **Iron Additives**: Iron can be applied to the soil as a granule or liquid. It can also be applied as a foliar spray, which is faster acting than soil applications.
- **Sulfur Additives**: Sulfur additives, whether granular or liquid, will acidify soils and unlock iron and zinc. They are faster acting than humus but do not provide additional health benefits.

TOO MANY NUTRIENTS

The soaps, oils, and organics in greywater contain nitrogen, phosphorus, and potassium—nutrients plants need to grow. But when an area receives a daily dose of this water, these nutrients can accumulate to toxic levels. The best remedy for overcoming nutrient loading is to use plants that can make use of it. Plant bigger plants, more aggressive plants, and/or more product producing plants.

Figure 10.8   **Green veins and yellow tissue are the signs of chlorosis and iron deficiency. This grape is infrequently irrigated with reclaimed water, which makes the soil alkaline and the iron unavailable. The remedies are to increase irrigation to move salts, apply compost or humus (not course mulch), or, and in worst case scenario, apply an iron additive, which can either be foliar or granular.**

OVERCOMING LOW PERMEABILITY/COMPACTION

Soil compaction is common in urban soils, and more so when the soil is constantly moist. Compaction reduces a landscape's ability to make the best use of greywater. Below are seven recommendations for overcoming compacted soils.

1. **Leave Weeds**: The plants that colonize compacted soils are early succession pioneers and help prepare the area for the later successions by loosening and enriching the soil. They increase permeability and reduce pooling. Do not pull weeds until a more favored plant is on hand to replace it.
2. **Redirect Traffic**: Keep machines and people traffic off moist and compacted soils.
3. **Aerate Soil**: Pulling plugs from soil will allow it to expand, which allows it to exchange its gases and create better growing conditions.
4. **Apply Mulch**: Dense, coarse, and recently grinded mulch will cushion soil.
5. **Sow a Cover Crop**: Cover crops are excellent at breaking up compacted soils and providing nutrients. As a rule, cover crops are comprised of grasses and annuals from the pea family, Refer to the Landscape Materials chapter for more detail.
6. **Thoroughly Dry Soil**: If possible, let the soil dry to at least 6 inches deep. Compacted soils tend to be expansive and will visibly rise and fall with wet and dry cycles. This physical action breaks bonds and increases porosity, increasing gas exchange and reducing compaction.
7. **Mechanically Turn Soil**: Turn over the soil and mix it with organic materials to encourage microbes, which create short-term airspace and porosity.

Occupant Behavior

Salts and solvents, tissues and wrappers, hair and sludge—a great number of unwanted things get flushed from laundry machines, showers and sinks. These unwanted items will impact the system and landscape health. Unfortunately, not everyone inside a building knows this.

Managing a healthy system means managing what goes down the drains, and that means managing occupant behavior. Several strategies will be needed to ensure a successful greywater system.

- **Signage** that explains etiquette is needed next to showers, sinks and drains.
- **Demonstrations** to employees and occupants may be required to showcase how to properly dispose of tissue, oils, and solvents.
- **Soaps** that are not harmful to plants and soil ecology are vital. See below.

GREYWATER COMPATIBLE SOAPS[8]

For the healthiest landscape, plants and occupants, the soaps that become greywater should have special characteristics. They should be biodegradable, non-toxic, neutral pH, and free of boron and salt. Avoid soaps that are anti-bacterial and have softeners, artificial fragrance, brighteners, or artificial colors.

Two Good Resources for Greywater Soaps

- Environmental Working Group, at https://www.ewg.org/guides/cleaners/content/top_products/
- American Cleaning Institute at https://www.cleaninginstitute.org/ingredientcommunication

Resources

BOOK

Ludwig. Art. *Create an Oasis with Greywater: Choosing, Building, and Using Greywater Systems, Includes Branched Drains.* Oasis Design. 2007.

WEBSITE

Greywater Action for a Sustainable Water Culture. Can be viewed at http://www.greywateraction.org/

Notes

1 Hard to escape the influence of Art Ludwig's work, intentional or not. The Landscape Characteristics are a composite of many sources and experiences, but Ludwig is woven though all of it.

Ludwig, Art. *Create an Oasis with Greywater: Choosing, Building, and Using Greywater Systems*. Oasis Design. 2009.

2 Sharvelle, Sybil, et al. "Long-Term Effects of Landscape Irrigation Using Household Graywater." *Urban Water Center*, Water Environment Research Foundation, 2006. https://greywateraction.org/wp-content/uploads/2014/12/Long-term-Study-on-Landscape-Irrigation-Using-Household-Graywater.pdf

3 California Plumbing Code (CPC) Title 24, Part 5, Chapter 16A, Part I – Nonpotable Water Reuse Systems. 2007 and 2009.

4 "Branched Drain Greywater System: Overview." Water Wise Supply, 2023. Accessed March 18, 2023. https://www.waterwisesupply.com/education/systems/branched-drain

5 California Plumbing Code (CPC) Title 24, Part 5, Chapter 16A, Part I – Nonpotable Water Reuse Systems. 2007 and 2009.

6 Carpenter, Susan. "Some Plants Thrive on Gray Water, Study Says." *Los Angeles Times, L.A. at Home*. October 3, 2012.

Sharvelle, Sybil, et al. "Long-Term Effects of Landscape Irrigation Using Household Graywater." *Urban Water Center*, Water Environment Research Foundation, 2006. https://greywateraction.org/wp-content/uploads/2014/12/Long-term-Study-on-Landscape-Irrigation-Using-Household-Graywater.pdf

Ludwig. *Create an Oasis with Greywater*. Oasis Design. 2009.

7 Goodridge, Lawrence, et al. "Evaluation of a Low-Cost Treatment System for Recycled Greywater use in Irrigation of Produce." *ASHS Annual Conference*, Colorado State University, 2011. Their research concluded that greywater run through a wetland will exit clean enough (free of bacteria) to be sprayed on lettuce.

8 This is a combination of work:

Stephen Murray (Culver City, CA) wrote extensively for the Sierra Club's Water Committee. June 2015.

Sharvelle, Sybil, et al. "Long-Term Effects of Landscape Irrigation Using Household Graywater." *Urban Water Center*, Water Environment Research Foundation, 2006. https://greywateraction.org/wp-content/uploads/2014/12/Long-term-Study-on-Landscape-Irrigation-Using-Household-Graywater.pdf

Chapter 11

Water: Rainwater Capture and Use

Capturing, storing, and utilizing rainwater is a linchpin. Water grows our gardens, and they grow our health and wellbeing. Rain is water at its most perfect. It is clean, it is free, and it is often abundant. The trick is being able to capture and utilize it. And if we can master that, we can grow healthy communities and people.

But capturing and utilizing rainwater does far more than just sustain our gardens.

- It helps protect America's waterbodies by reducing the polluted runoff flowing from parking lots, rooftops, and streets.
- It helps protect a region's stormwater drainage system from being overwhelmed and reduces costly maintenance and upgrades.
- It helps replenish dwindling aquifers and supports the sustainable management of ground water.

Covered below are the two types of rainwater capture: Passive and active. Passive capture means that the rain is led into a landscape and encouraged to infiltrate. Active capture involves storing the rainfall for later use.

Rain Gardens (Passive Capture)

Rain gardens capture water passively by storing it in the ground. Rain gardens are not only fantastic for landscape health and recharging aquifers; they also protect the state's waterbodies.

Our Nation's waterbodies—our creeks, streams and ocean—are under assault. Biodiversity has plummeted, the quality of drinking and recreational waters has declined, and commercial and sport fishing stock has fallen. Most of the pollution driving those changes is stormwater runoff from urbanized areas. Rain gardens can greatly reduce stormwater runoff. But reducing urban runoff is no easy task. With up to 70% of urban

DOI: 10.4324/9781003369752-11

Figure 11.1 **This chapter covers two aspects of rainwater capture: Passive and active. Passive methods are often referred to as rain gardens. Active capture involves collecting rain in barrels, cisterns, and tanks for later use.**

areas under impervious cover and up to 50% of all rainwater falling turning into runoff, managing these vast, fast and sometimes furious flows is difficult.

The job of a rain garden is to slow, stop and clean runoff. Its purpose is to reduce the speed at which runoff flows, provide opportunities for runoff to infiltrate, and to screen the water that does run off. The particulars of this job are below.

Rain Garden Overview

Uses: A property's rainwater is kept onsite and allowed to infiltrate with the goals of increasing soil moisture, growing a greater range of plants, recharging an aquifer, reducing stormwater, and protecting downstream environments.

Costs: Low to medium. Between 2007 and 2008 the Orange County Coastkeeper (CA) found that homeowners were able to divert their roof water into their landscape for less than $500.

Difficulty: Not difficult to understand or construct, but a good understanding of hydrology and managing moving water is needed.

Best Scale: Passively capturing rainwater works at every scale.

ERoEI: Infiltrating runoff onsite helps create healthy communities, but that does not mean all those benefits are experienced onsite. There are costs and risks with managing

Figure 11.2   **Stormwater management can be simple. Pictured is a grassy depression that captures and helps clean runoff from a parking lot. Without this design the polluted runoff would flow to a river that empties into the Pacific Ocean. Picture taken in Santa Ana, CA.**

stormwater. While the ERoEIs of a rain garden may not be beneficial for the property manager, they help the community with rising ground water, better recreational opportunities, and more biological diversity.

Nomenclature

Aquifer: An underground formation capable of storing and providing water.

Bioretention: A planted area designed to collect, filter, and/or allow water to infiltrate.

Catch basin: A belowground receptacle that allows debris and sediment to settle out of runoff before the water is diverted to a drainage system.

Detention Basin: Any pond or pool that holds water for later release. Detention basins help reduce peak discharge during a storm, preventing a stormwater system from becoming overwhelmed in high rain events.

Dry Season Runoff: Surface runoff that is not generated by rainfall. Common sources of dry season runoff include over-irrigation, emptying pools, and washing urban surfaces.

Hydrograph: A graph showing the amount of water flowing past a specific point in time and place, such as a channel, creek or river. Urban hydrographs are far different than woodland hydrographs. Urban areas discharge water quickly, whereas woodland areas do not.

Impermeable: A material that does not allow water to pass through it, which can be naturally occurring or manmade.

Infiltration: The flow of water downward, from a land surface to a subsurface.

Nonpoint source pollution: Pollution coming from a large area, but not a specific source. Residential and commercial properties are considered nonpoint sources of stormwater pollution.

Permeable: A material that allows water to pass through it.

Point source pollution: Pollution that has an identifiable source, such as a specific pipe, factory, or land use, like agriculture.

Retention basin: Any pond or pool that holds water. The reason to hold the water may be for infiltration or it may be to delay release, helping to reduce a peak discharge (called a detention basin).

Runoff: Water that runs over the top of a surface.

Screening: A device or process that screen debris from runoff. Common devices include catch basins and swales.

Stormwater runoff: Water running over a surface and generated from rainfall.

Swale: A constructed and vegetated channel used to direct, slow, filter, and infiltrate runoff.

Waterbody: A body of water, which includes aquifers, creeks, streams, rivers, lakes, ponds, and oceans.

Best and Worst Types of Properties for Passive Rainwater Capture

BEST PROPERTIES

There are many types of properties where diverting the site's runoff into the landscape is a great idea.

Over an Aquifer: Many communities sit on top of their source of water and many of these aquifers are overdrawn. Sinking rainfall and runoff into a landscape allows more of that water to make it to the aquifer, supporting a local resource.

Permeable Soil: Any property with permeable soil is a good candidate for a rain garden. Good soil allows for good rates of infiltration, which thwarts many of the problems with rain gardens, such as erosion, pooling, and vector problems.

Infrequent Rainfall: Properties in areas with infrequent rainfall can greatly enhance landscape health by storing rainfall in the landscape. Properties that allow rain to infiltrate take longer to dry, support good soil health, and will improve plant health.

Waterbody Protection: Americans love our lakes, streams and oceans. The surest way for any individual to protect these cherished waterbodies is creating a property that prevents polluted and toxic runoff from entering those aquatic ecosystems.

WORST PROPERTIES

Not every property should hold on to their rainfall and runoff because of the risk the water poses to the site and/or downstream development and/or ecology. These properties are:

Hillsides/Slopes: Infiltrating water on sloped properties can be dangerous. Debris flows, landslides and soil slips become more likely as soil saturation increases. Dry slopes are more stable than wet ones. Infiltrating large amounts of runoff on a sloped property should involve an Erosion Control Specialist.

Dense Soil: Properties on bedrock or clay struggle to properly infiltrate rainfall and runoff. Legally, compounded water must be gone within 72 hours, any longer and the water technically becomes a pool, which has a completely different set of laws. Unwanted pooling can lead to vector problems, the concentration of toxins, and public health hazards.

Toxins: Any property with a history of manufacturing, storing and/or shipping hazardous materials is problematic. On one hand, the site should not infiltrate rainfall and push the toxins towards an aquifer. On the other hand, discharging runoff offsite puts the chemicals in the storm drain system and eventually a waterbody. The ideal strategy is to biologically clean the runoff through a constructed wetland.

Lowest Point: When an asset, such as a house, business or outbuilding, sits at the lowest point of a property, it is always at risk of flooding during rain. Jettisoning runoff from these properties is vital in protecting these assets.

Figure 11.3   **Infiltrating large amounts of water on slopes is ill-advised. The seepage will increase the chances of soil slips and soil slumps. Runoff from impervious surfaces should be directed into a stormwater system.**

Passive Harvest: Designing a Rain Garden

A waterbody protecting landscape is easy to understand and design. There are only three goals: Slow, stop, and clean runoff.

SLOWING RUNOFF

Slowing runoff helps protect a waterbody for three reasons. First, it reduces the speed of runoff and consequently, the amount of damage it can do. Second, slowing water allows debris and sediment to settle out. And third, slowing water provides a longer opportunity for infiltration. Below are some of the many strategies to slow runoff.

Curb Cuts: Cutting into curbs allows runoff to flow into landscaped areas. Studies have shown that curb cuts can help stop all dry season runoff and the runoff from small storms. Typically, between 6" and 1', curb cuts are inexpensive and are ideal for parking lots, roadways, and walkways.

Figure 11.4 **A curb cut allows the water running in a gutter to flow into a planted area. They are simple and effective. Pictured is a curb cut in a parking lot medium.**

Breaking Up Impervious Areas: Disconnecting large areas of impervious material is critical to slowing/stopping runoff. Breaking up these areas without losing their functional qualities can be accomplished with decomposed granite, drain grates, gravel, pavers, and wooden decks.

Evergreen Trees: All trees intercept rain. A deciduous one can diffuse up to 760 gallons per year, a mature evergreen up to 4,000 gallons.[1] Studies estimate that urban forests can reduce annual runoff by as much as 7%, but it could be as high as 12% with more urban evergreen trees.[2] Deciduous trees not only intercept less rain, but also litter profusely; some as much as 200 pounds of leaf litter a year. Deciduous is not recommended near impervious surfaces that lead to storm drains, such as parking lots and streets.

Many Vegetative Layers: A landscape with many layers of plants, such as trees, shrubs and ground covers, is fantastic at diffusing, absorbing and slowing rainwater. When a dense forested landscape in North Carolina was converted to turf the rate of rainwater infiltration fell from 12.4 inches an hour to 4.4 inches an hour.[3]

Mulch: All mulches slow runoff, increase infiltration, and reduce evaporation. Although organic mulches are preferred for healthy productive landscapes, inorganic mulches are ideal for windy locations and high traffic areas.

Undulating Surfaces: The smoother the surface, the faster water runs. Creating landscapes, whether planted or not, with undulating surfaces puts obstacles in the path of runoff and slows it.

Permeable Materials: The best type of permeable surface depends on its level of use. High use areas require the safest surfaces, which are usually the least permeable. Conversely, low use areas can afford to use the most permeable materials, like mulches. Use the chart below to select the best materials for every level of use.

Figure 11.5   **Deciduous trees should not be planted along streets. Not only do they drop hundreds of pounds of leaf litter a year just before the wet season and clog or hinder storm drain systems, but they also do a poor job of protecting a community from the impacts of rain and snow.**

Table 11.1:   A Comparison of Permeable Walking Surfaces

Surface	Good Uses & Notes	Infiltration/Runoff	Costs
Brick	Closely spaced brick can be used for all walkways; loose bricks are good for less frequented paths.	Low to high, depending on size of gaps between bricks.	Moderate to high installation costs; low to moderate maintenance costs.
Crushed Aggregate	Crushed aggregate is a common, all-purpose surface. Pea gravel is better on bare feet but is not as porous.	High.	Low maintenance and installation costs. Weeds are common.
Decking	Decking is used for all levels of use if the gaps between boards are small.	Moderate to high, depending on gaps between planks and the surface below the deck.	High installation and maintenance costs. Should budget for replacement every 10–15 years.

(Continued)

Table 11.1:   Continued

Surface	Good Uses & Notes	Infiltration/Runoff	Costs
Decomposed Granite (dg)	Excellent for garden paths. Keep away from building entrances because it sticks to feet and gets carried in.	Moderate to high. Permeability is related to degree of compactness.	Low installation and maintenance costs, but weeds are a problem.
Greenroofs	A roof that is extensively planted can capture up to 75% of a small rain event and slow rainwater in large ones.	Low to high, depending on amount of rain and depth of soil medium.	Very high installation and high maintenance costs.
Organic Mulches: Bark, grindings, hulls, shaving.	Comfy walking surface in shoes, nutrient providing, and slows soil evaporation.	High.	Low installation and maintenance costs, although weeding will be necessary. Apply biannually.
Pavers	Bricks and pavers provide a safe walking surface.	Low to high, depending on gaps and degree of slope.	Moderate installation and low maintenance costs.
Porous Asphalt	Good replacement for current asphalt driveways. However, it is petroleum based and produces dirty runoff.	Low to moderate.	High installation and low maintenance costs. Can lose 75% efficiency in 5 years. Vacuuming helps maintain permeability.
Porous Concrete	Good anywhere around a structure and light-duty driving. It does not hold up as well with frequent vehicle traffic.	Low to moderate.	High installation and low maintenance costs. Can lose 75% efficiency in 5 years.[14] Vacuuming helps maintain permeability.
Flat Stone: Natural, Recycled, or Manufactured	Excellent for paths and patios, but not ideal for high-heels and high use areas.	Moderate to high, depending on size and gaps.	Low to high installation costs, depending on the type of material. Low to moderate maintenance costs, depending on the types of plants and materials filling gaps.
Turf Block	Great for short term parking areas. Tough to properly irrigate if cars are always on it.	Moderate to high.	Moderate installation costs and moderate to high maintenance costs, more so if the grass must be mowed and irrigated.

Figure 11.6   **The developers of this condo complex went to great lengths to slow runoff and provide opportunities for infiltration. Pictured are colored porous concrete and pavers.**

STOPPING RUNOFF

Stopping runoff and providing opportunities for infiltration is vital to waterbody protection. There are two types of infiltration devices: above and below ground.

ONLY INFILTRATE RAINWATER IF ....[4]

Not every property should infiltrate rainfall and runoff. A good property will have the following attributes:

- No history of the manufacture, storage, or handling of toxins such as paints, petroleum products, and solvents.
- Infiltration area is at least 20' downhill of a structure or 100' uphill.
- The infiltration area is on a slope no greater than 20%.
- A location that does not have a slope that is 35% or more downhill from the infiltration area.
- The rate of infiltration is no less than .5" per hour.
- The infiltration area is easy to access and maintain.
- The infiltration area has an emergency overflow device/method.

Testing the permeability of soil is essential in determining the feasibility of an infiltration system. While large developments are mandated to have this test done by a lab, smaller properties can employ a lower-tech method:

1. Dig a hole that is 3' by 3' by 3' and fill it with water.

2. After all the water recedes, fill the hole again and start a timer.
3. If the water is not gone in 72 hours, then the site is no good for infiltration. If the water is gone within 48 hours, it is okay for infiltration. And if the water is gone in less than 48 hours the soil is good for infiltration.

Above Ground

Above ground infiltration devices are preferred over below ground. Above ground devices are quickly and inexpensively constructed. They are also easily monitored, altered and maintained. Below are the most common above ground devices.

Buffer Strips/Filter Strips: A combinations of trees, shrubs, and/or ground covers that parallel a road or other source of polluted runoff. Shelter strips protect streams, shorelines, and storm drains from pronounced sources of pollution. These depressed vegetative strips will not only provide a biological block to polluted runoff but filter the air as well.

Dry Stacked/Stepped Walls: Dry stack walls are a quick and efficient way to stop runoff on low to medium steep slopes. They are small and sometimes long walls running mostly perpendicular to a slope or water path.

Infiltration Basin/Seasonal Pond/Stormwater Pond: Any place where water is allowed to sit and infiltrate. Plants are highly recommended for these permeable areas. Rates of infiltration tend to fall in unplanted basins.[5] Refer to the section on Designing an Infiltration Basin below for greater detail.

Micro Basins: Micro basins can be used throughout a landscape and are effective on flat to medium steep slopes. Micro basins are simply small depressions dug into a landscape.

Swales: Swales are constructed to help move runoff through a landscape. Swales are earthen, or have earthen bottoms, and allow for infiltration. A grass and rock lined swale can stop up to 85% of the runoff from a small to medium rain event.[6]

Vegetative Cover: Dense vegetation is one of the most effective methods for slowing and stopping rain from becoming runoff. A richly forested landscape, for example, can absorb up to 70% of a medium rain event. Vigorous plants also mean soils rich with roots and life, both of which significantly increase porosity. As opposed to lawns, between to 30% to 50% more rain is captured in landscapes planted with trees, shrubs, and ground covers.[7]

Designing an Infiltration Area

An efficient infiltration area will have these eight characteristics:

1. **Away from Structure**: Water should never be allowed to infiltrate next to urban assets, such a building. At a minimum, water should be led to an area that is no less than 10 feet from a structure.

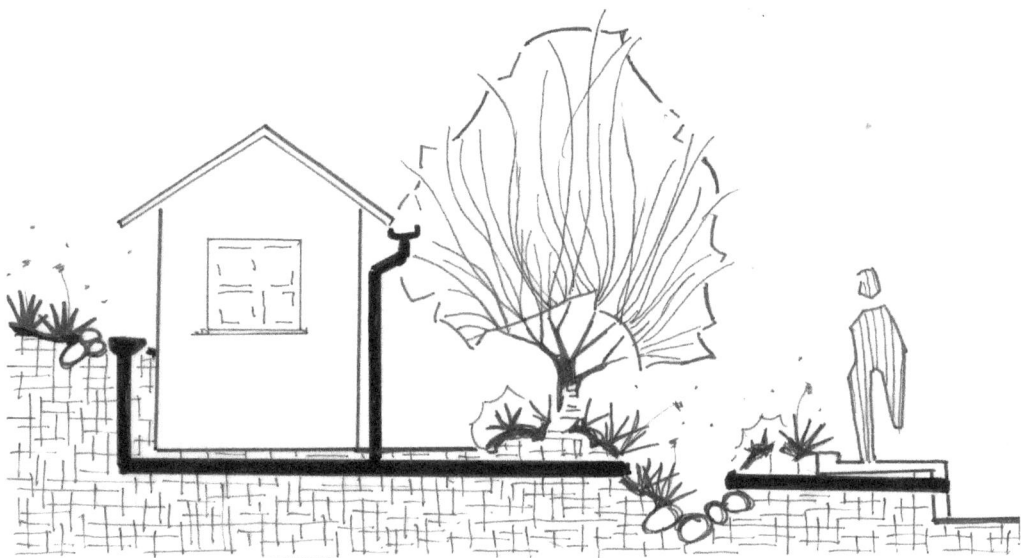

Figure 11.7   An efficient infiltration basin will have certain attributes: A location away from a structure, the right size, good soil, the right configuration, good access, plenty of plants, regular cleaning, and an emergency exit for overflowing water.

Table 11.2:   Sizing Infiltration Basins Based on Land Use

Land Use	Percent (%)
Completely Paved	4
Urban Freeways	3.5
Industrial Areas	3
Commercial Areas	2.5
Institutional Areas	2
Construction Sites	2
Residential Areas	1.5
Open Space	1

2. **Right Size**: Infiltration basins are 3% to 4% of the area draining to it. These estimates include the related landscaping, such as filter strips, fencing and paths. The chart below offers an approximate size for infiltration basins based on land use. A professional is required for large and/or public projects[8].

   The amount of space can also be calculated using the mass of water. For example, if the amount of water expected can be calculated, then multiply it by .13368 to get the space it will occupy in cubic feet. If size of the infiltration is known (in cubic feet), then multiply it by 7.4805 to get the number of gallons it will hold. For instance, 161 cubic feet of space can accommodate 1,204 gallons of water.

3. **Good Soil**: A good infiltration basin will have soils that can absorb at least .52" of water per hour,[9] which is about 33 gallons per 100 square feet. Ideally, the soil will be less than 30% clay. Non-residential projects will need to have the soil professionally tested.

4. **The Right Configuration**: The design capacity of an infiltration basin does not change, but its configuration will, depending on its type of soil. Clay soils

demand a shallow and large basin; water levels should never be no more than 6" deep. Sandy soils, on the other hand, can be up to 2' deep and occupy a smaller area.

5. **Good Access**: Construction and maintenance practices will compact soil. Strive to design an infiltration area that can be built and maintained from its perimeter, rather than its interior.

6. **Plenty of Plants**: Plants can both help and hinder infiltration. Vigorously growing plants will rapidly take up water creating pore space that increases infiltration. These same plants can quickly colonize an area and reduce the size of the basin and the amount of water it can hold. Although plants, and all the life associated with them, are the keys to cleaning runoff and increasing infiltration, they will have to be regularly thinned to maintain the design capacity of the basin.

7. **Regular Cleaning**: Infiltration basins are designed for specific volumes of runoff. As an infiltration slowly fills with plants, debris and sediment, the design capacity is reduced. Periodic cleaning is needed. As a rule, infiltration basins are cleaned when they lose 10% of their capacity.

8. **Emergency Exit**: A infiltration area must be designed with an emergency exit for the water. During periods of heavy rainfall infiltration basins can become overwhelmed. Instead of allowing the water to back up and endanger urban assets, the water should be jettisoned offsite and into the storm drain system.

Below Ground Infiltration Devices

Below ground devices are used frequently in densely urban areas. They are perfect where space is limited, like between two large buildings. These devices are more expensive to construct and maintain. They also lose efficiency rapidly if debris and sediment are not screened from the runoff. Below ground infiltration devices work best with cleaner runoff, like roof water, as opposed to street water.

Dry/Infiltration Well: A dry well is a large pit lined with filter fabric and filled with a variety of porous objects, like aggregate, milk crates, or plastic boxes. Dry wells are generally no smaller than 3' × 3' × 5' and are typically deeper than wide. Large wells will include an inspection port where rates of infiltration can be visually measured.

French Drain/Infiltration Trench/Recharge Trench: A long fabric lined trench filled with rounded aggregate/gravel. French drains are good devices to use in narrow areas. Unfortunately, if not installed or maintained properly, these devices have a 50% failure rate after just five years.[10]

Soakaway: A large concrete lined pit with earthen bottom. Soakaways are commonly used in dense urban areas. These pits are typically covered with metal grates and filled with large rocks. A 60 cubic feet soakaway (about 6' deep, 5' long and 2' wide) can accommodate a 2,200 square foot drainage area.

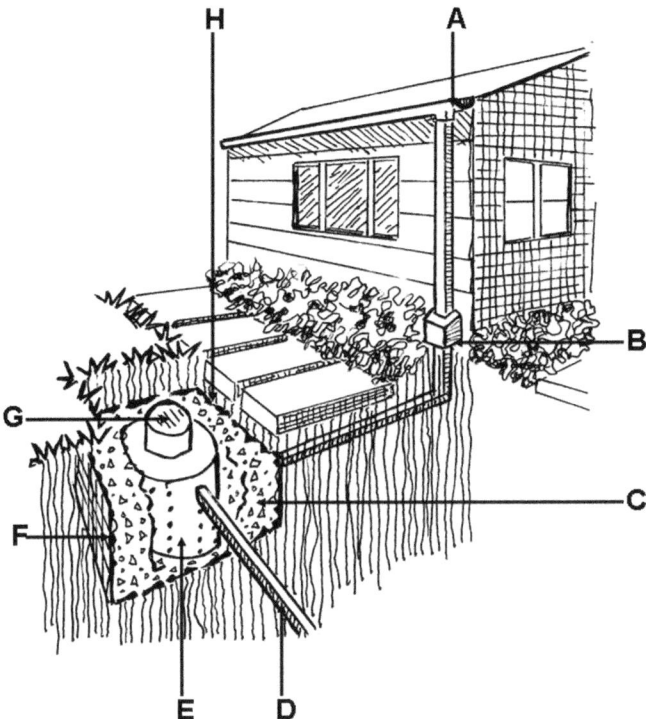

Figure 11.8   **Highlighted above are the design elements of a below ground infiltration device. A. Leaf guard in rain gutter B. ¼" Screen on downspout C. The aggregate is rounded, 1" to 1.5" diameter, and thoroughly washed D. Overflow to storm drain. E. Any assortment of devices, or no less than a 2" perforated drainpipe F. Commercial grade sediment/filter cloth, but not on the bottom G. Inspection and cleanout access H. Dry well not less than 10' from a structure**

SCREENING/CLEANING RUNOFF

As discussed above, infiltrating rainfall and runoff is not for every property. In these situations, screening/cleaning becomes the goal. If a property cannot reduce the volume of runoff, then at least it should try to improve the quality of runoff it produces.

Screening runoff is the minimum goal for waterbody protection. Screening removes suspended material, like leaves and sediment. Cleaning removes everything else; metals, nutrients and pathogens are taken out of runoff using biological processes. Cleaning runoff requires far more time and effort.

No matter the type of device used to improve the quality of runoff; all should have these characteristics:

- They should have an overflow outlet. Every device has the potential to flood in severe rain.
- They should be the last device runoff flows into before leaving a property.
- And all these devices should have good accessibility because they will require regular maintenance.

Figure 11.9   **Maintenance is essential. Poorly maintained stormwater systems can lead to erosion, pooling, and the degradation of urban infrastructure. The grass blocking this curb cut is a great example. Moisture is pooling in the parking lot, not the swale.**

Screening Runoff

If a property cannot hold its runoff, then screening it is the very least the landscape should be designed to do. Below are the commonly used devices.

Bio-Catch Basin: A large depression where runoff is led. They are densely planted and trap plant litter and sediment.

Catch Basin: Any device that allows the heavier material to settle. They come in all types of sizes and designs and are common in urban areas.

Detention Basins: A detention basin impounds water and then slowly releases it after peak flow has passed. They are used for flood control and sediment removal and are designed to handle large volumes of runoff.

Fiber Rolls: Fiber rolls are straw, rice hulls or coconut waste bound by strong plastic mesh. They are used around storm drains and on slopes. They are highly permeable and allow sheeting water through but trap debris. They will become a litter nuisance if not removed after one year.

Filter Cloth: Fabrics are not only used to filter solids out of runoff, but particles out of the air, too. They are a common feature on construction sites, where fabric may surround a property and lay over storm drains.

Gutter Diffusion Box: Like a catch basin, a gutter box removes large debris. These devices, however, are specifically designed for the large and fast flow of roof runoff. They are common in wooded communities because of the high levels of leaf litter.

Sediment Pond/Trap: A widening in a channel that allows the water to slow enough to drop its heavier particles. These areas can be permeable or not, vegetated or not.

Swale: A swale is used to channel water and encourage infiltration. Technically, a swale is no more than a large ditch. It can be vegetated or not, rock lined, or not. A bio-swale is a planted swale. A dry creek is a swale lined with rocks.

Cleaning Runoff

Stormwater is cleaned via a constructed wetland. Allowing stormwater to slowly move through the root zone of a healthy wetland will help remove chromium, fertilizers, hydrocarbons, iron, lead, mercury, oil, pathogens, pesticides and solvents. Runoff is cleaned through the action of bacteria, bugs, fungus, and the uptake of plants.

Figure 11.10   **Pictured is a constructed wetland that detains and cleans water flowing from a 13.2-acre commercial development along the coast. Without this type of land use, the stormwater flowing from the buildings and parking lots would flow directly into the ocean. The wetland consumes about 1 acre and is heavily planted with bulrush and sedge. Picture taken in Newport Beach, CA.**

Structural Design of a Constructed Wetland

The Bottom: The bottom of a constructed wetland can be permeable or not. However, impermeable bottoms are sometimes more effective because the longer stormwater is in a wetland, the cleaner it becomes.

Overflow Outlet: Any depression that holds water may flood and overflow during periods of intense rain. To ensure that this overflow does not endanger the property or downstream environments, an overflow outlet is needed to direct the flooding water into a storm drain system.

Good Access: Constructed wetlands need regular maintenance. Access via truck or handcart is vital to make the maintenance as easy as possible.

Plants: The types of plants used in constructed wetlands are unique. They tolerate saturation; they have low sensitivity to metals, salts and toxins; and they are often native. Some of the best plants for wetlands include:

Alder, *Alnus* spp.
Arundo (carrizo), *Arundo donax*
Bur marigold, *Bidens laevis*
Cattail, *Typha* spp.
Comfrey, *Symphytum* spp.
Aspen, Cottonwood and Poplar, *Populus* spp.
Bulrush, *Bolboschoenus* spp. and *Schoenoplectus* spp.
Horsetail, *Equisetum* spp.
Iris, bog type
Rush, *Juncus* spp.
Papyrus, *Cyperus papyrus*
Plane Tree and Sycamore, *Platanus* spp.
Sedge, *Carex* spp.
Willow, *Salix* spp.

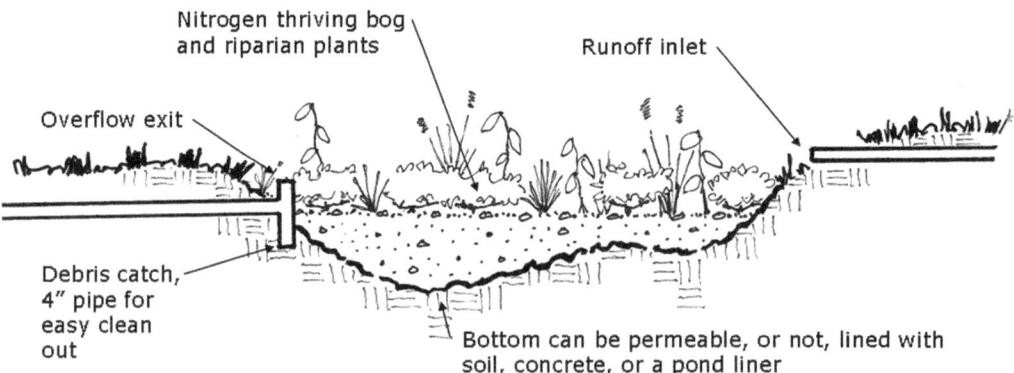

Constructed Wetlands

Figure 11.11 **The design attributes of a healthy constructed wetland.**

Maintenance

The maintenance of a biological cleaning device is important to note. While many of the toxins may have been removed from the runoff, they are still in the environment, locked in soils and plant tissue. These harmful pollutants, mainly metals, still need to be disposed in a place where they will not get back to the environment, such as a landfill specializing in toxic materials.

Active Harvest: Rainwater Storage

Capturing rainwater is an art and practice thousands of years old. Rainwater is prized because of its soft nature. It extends the life of appliances and is much better for any type of cleaning. Rain starts with zero bacteria, chemicals, minerals and salts, although it begins to lose its purity as it falls through urban atmospheres.

For brevity, this section has a sharpened focus. It assumes the rainwater collected and stored is used in a landscape, that the water is non-potable, and that the collection surface is unintentional, such as roofs or patios. Rain Barns and other intentional rain harvesting devices are not covered in this book.

Active Capture Overview[11]

Uses: The rainwater is collected and used to grow various landscape goods and services.

Costs: Low to high, depending on the type of system.

Difficulty: Not difficult to moderately difficult. Expertise needed includes hydrology, plumbing, and general building.

Best Scale: The best site for a rainwater system is a property that genuinely needs more irrigation water. The site may have water restrictions, or not be connected to a municipal supply, or be on a limited supply of ground water. Because of the embodied costs, a rain garden is preferred over a capture system on properties with less than 1,000 gallons of captured water.

ERoEI: Importing potable water into urban areas has an energy cost of 15 BTUs to 30 BTUs per gallon of water. The energy footprint of a rainwater capture system, which includes the installation and maintenance of rain gutters, plumbing, tanks, and irrigation, can exceed 70 BTUs per gallon.[12] The trick with capturing rainwater is to make sure that the cost—the plumbing, plastics and fossil fuels—creates a dividend that exceeds the costs of capture. Some of these dividends include cooling, food production and/or personal health.

Nomenclature

Rainwater falls on a catchment area, is screened of large debris, and then is transported to a tank for storage. Once in the tank, sediment is removed from the water and if not gravity driven, then it is pumped into the landscape. Below are the basic components and nomenclature of a capture system.

Capture Surface: Any surface that allows water to sheet can be used to capture rain. These surfaces can be elevated or not. Elevated surfaces, like roofs, are preferred because of the gravitational energy embodied with height. Surfaces can also be intentional or not. Intentional surfaces include Rain Barns and fog and mist nets. Unintentional surfaces are roofs, patios, and solar panels.

Transport System: A system that collects water from the catchment area and leads it to storage. Ground transport devices include drainpipes, gutters, culverts, and swales. Above ground devices include roof gutters, downspouts, rigid drainpipes, and sluices.

Screening Devices: Screening the debris in runoff before storing it improves the quality of the stored water and helps reduce plumbing, storage, and pumping failures.

First flush: Directing the first few rains of a season into a landscape, rather than into a collection device, greatly enhances the efficiency of a storage system. The first few rains of a season are loaded with debris and toxins.

Storage: Rainwater can be stored in an array of devices and materials, with capacities varying from 55 gallons to thousands of gallons. Some of the most common storage devices include barrels, cisterns, ponds, and tanks.

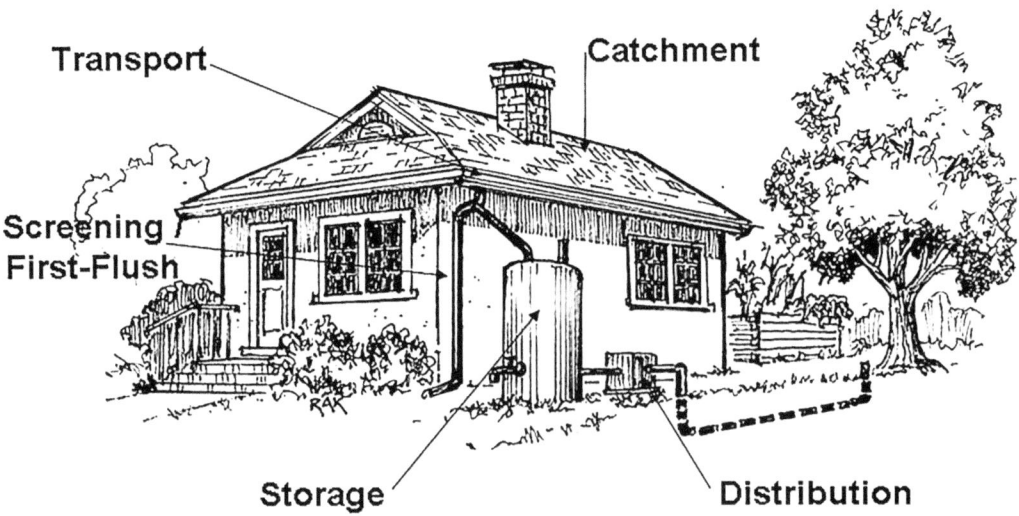

Figure 11.12   Illustrated above is an overview of actively capturing rainwater for future use.

Treatment: Creating potable water from rainwater is rarely done on small sites because the costs are too great. The techniques and technology used to make water potable include chlorination, ozone, reverse osmosis, and UV light.

Distribution System: Moving stored water to its intended use can be low-tech or high-tech. Low-tech systems are gravity fed and used for landscape irrigation. High tech systems involve pumps and produce significant water pressure with many uses.

Best and Worst Types of Properties for Active Rainwater Capture

BEST PROPERTIES

Genuine Need: The best site for a rainwater capture system is a property that genuinely needs more irrigation, be that for landscape health, food production, or fire protection.

Ease of Use: An ideal property will make the use of rainwater easy. Typically, that means that the water is elevated and slightly pressurized, which makes it easier to move and utilize.

Involved Users: Whether or not a system provides benefits to the environment and site hinges on how it is employed—the energies of a committed group of people are what define the success of a system, not the system itself.

WORST PROPERTIES

Inadequate Maintenance: Storing rainwater requires labor. A property is not suitable if it cannot guarantee a basic level of maintenance:

- The water must be used—and the sooner the better—it provides more opportunities for refill.
- Rain gutters and water screens need to be regularly cleaned, both before the start of the rain and at least once during the wet season.
- Tanks will need to be cleaned of debris every other year.
- And diligence is always required to ensure the water is inaccessible to wildlife, most notably mosquitoes and varmints.

Benefit/Cost: Capturing and storing rainwater has costs. Whether that cost is environmental and related to amounts of plastics and energy use, or personal and related to expense and time, a water capture system should produce a benefit that exceeds the costs. Many properties cannot produce more benefit than cost.

Designing for a Need

Designing a rainwater harvest system starts with identifying need. There are three primary reasons to capture rainwater in urban areas: Conservation, fire protection, and reliability.

Conservation: Civic minded people harvest rainwater to supplement the water they get from a municipal/centralized supply. Since the need is less of a matter of sustenance, any amount of harvesting is assumed to be beneficial. Typically, storage devices vary between 55-gallon barrels to 5,000-gallon cisterns.

Fire Protection: There are two aspects to watering for fire protection. First, soaking the landscape before a fire. This is preventative and helps protect a landscape from firebrands and flame contact. Second, using the stored water to fight a fire. This requires a large amount of highly pressurized water.

Water Reliability: The capture system is built around on occupant and landscape water needs and the duration for which that water is needed. A system can be constructed for weeks or months.

Figure 11.13   **This 250-gallon rainwater storage tank helps support this community garden during drought restrictions. It supplies water to the neediest plants, such as those in pots, under trees, and ones that produce food. Picture taken at Arlington Garden, Pasadena, CA.**

SUMMARY OF SYSTEMS

Table 11.3: Determining Amount of Rainwater to Store Based on Occupant Need

	Ideal Storage Size, Gallons	Type of Water	Catchment Area Size	Ideal Pressure
Conservation	55 onwards	Non-potable	Any size	Gravity
Fire Protection	2,000 min. 3,000 is better	Non-potable, but must be free of pipe clogging debris	400 to 1,000 sq. ft.	20psi (soak) to 100psi (fight)
Reliability	2,000 to 10,000	Potable	400 to 2,000	30 to 80psi to run most indoor appliances

System Components

CATCHMENT AREA

Any surface that produces runoff can be used to collect rainwater. Catchment areas can either be intentional, or not. Intentional catchment areas are structures built specifically to capture rain, such as rain barns and mist nets. Unintentional catchment areas are an urban area's impermeable surfaces, the most common being the roofs of buildings.

HEIGHT

Surfaces should be elevated because the water is easier to screen and move. The greater the height, the greater the potential energy and uses. Not only is ground water harder to capture, but it is dirtier, too.

SURFACE TYPE

The ideal surface is slick and non-toxic. Glass is the best surface. Other good roofing materials include concrete, plexiglass (greenhouses), slate, stainless steel, and tile.

Materials to Avoid

Landscape materials get battered and a wide variety of toxic chemicals are used to preserve them. These toxins are found in color coatings, fasteners, flashing, lead solder, and wood. Foremost, avoid petroleum-based surfaces, like asphalt, tar and gravel. Also avoid any surface treated with a preservative, such as wood shingles. And lastly, avoid rough roof surfaces, like shingles and terra cotta, because they trap particulates and dust, and are more likely to support algae, bacteria and fungi.

CALCULATING AMOUNT OF HARVESTABLE WATER[13]

Calculating the number of gallons a surface provides is simple. The formula is square feet of surface area × .55 × amount of rainfall in inches.

It is important to note that the actual figure for gallons of water per square foot per 1" of rain is .627. The reason the equation uses .55 is because of loss. Common causes for losses are evaporation, faulty fittings, leaks, and blocked gutters and screens.

For example, with an average annual rainfall of 22" on a 1,200 square foot roof will produce 14.520 gallons of harvestable rainfall a year: 1,200 × .55 × 22.

TRANSPORT SYSTEM

The transport system directs water from a collection surface and deposits it into a storage device. The devices employed are rain gutters, downspouts, and drainpipes.

Gutters and downspouts are made from a variety of materials, whereas drainpipe is typically PVC. Gutters may be aluminum, copper (not recommended), galvanized steel, plastic, PVC and vinyl. Urban rainfall is slightly acidic and copper degrades in acidity, producing runoff that is toxic.

Gutters must have an incline of 2% to efficiently move water. No less that a ¼ inch fall per 10 feet is required; a ½ inch fall over 10 feet is recommended.

SCREENING

Screening is important: The cleaner the water is when it goes in, the cleaner it is when it comes out. The three most common screen are: Leaf guards that sit over rain gutters; downspout screens that sit between the downspout and gutters; and strainers that that sit just before the downspout empties into the storage device.

Importantly, never capture the first few rains of a season. Roofs and gutters are covered in a toxic mix of fine dust and toxins and storing those pollutants can create even more toxins. Store clean water only.

STORAGE

Some of the devices used to hold rainwater include barrels, cisterns, ponds, pools and tanks. Water is heavy and any above ground storage device holding over 100 gallons requires a fortified footing with good drainage immediately surrounding it. Maintenance of water storage systems is crucial, as lapses lead to algae, mosquitos, septic water, and unwanted erosion/pooling.

DISTRIBUTION SYSTEM

A distribution system moves water from storage to its intended purpose. Moving water with gravity is always preferred. Devices for gravity driven irrigation include large diameter hoses, trenches and flooding. Firefighting and other high-pressure endeavors require large pumps, pressure tanks, and thick hoses.

Resources

BOOKS

Banks, Suzy and Heinichen, Richard. *Rainwater Collection for the Mechanically Challenged*. Tank Town Publishing. 2004.

Kent, Douglas. *Ocean Friendly Gardens: A How-To Gardening Guide to Help Restore a Healthy Coast and Ocean*. Surfrider Foundation. 2009.

Lancaster, Brad. *Rainwater Harvesting for Drylands and Beyond: Guiding Principles to Welcome Rain into Your Life and Landscape*. 3rd ed. Rainsource Press. 2019.

Ludwig, Art. *Water Storage: Tanks, Cisterns, Aquifers, and Ponds for Domestic Supply, Fire and Emergency Use—Includes How to Make Ferrocement Water Tanks*. Second Edition. Oasis Design. 2005.

WEBSITES

A Beginner's Guide or Rainwater Harvesting. Tree Huggers. Can be viewed at https://www.treehugger.com/beginners-guide-to-rainwater-harvesting-5089884

American Rainwater Catchment Systems Association (ARCSA). Can be viewed at www.arcsa.org

Stormwater Management Practices at EPA Facilities. Environmental Protection Agency. Can be viewed at https://www.epa.gov/greeningepa/stormwater-management-practices-epa-facilities

Stormwater Program. California State Water Resources Control Board. Can be viewed at Storm Water Program | California State Water Resources Control Board

Turn Scarcity into Abundance. Rainwater Harvesting for Dryland & Beyond. Can be viewed at https://www.harvestingrainwater.com/

Notes

1   Cotrone, Vincent. "The Role of Trees & Forests in Healthy Watersheds." *Forest Stewardship Bulletin*, vol. 10, Penn State University, Extension. August 30, 2022. https://extension.psu.edu/the-role-of-trees-and-forests-in-healthy-watersheds#:~:text=Managing%20stormwater%2C%20reducing%20flooding%2C%20and%20improving%20water%20quality.

2 "How Trees Can Retain Stormwater Runoff." Edited by Dr. James R. Fazio. *Tree City USA Bulletin No. 55.* The Arbor Foundation, Nebraska City, NE. No date.

3 Cotrone. "The Role of Trees & Forests in Healthy Watersheds." 2022.

4 Kent, Douglas. *Ocean Friendly Gardens: A How-To Gardening Guide to Help Restore a Healthy Coast and Ocean.* Ed. Geever, Joe. The Surfrider Foundation. 2009.

5 1995. *Environmental Hydrology.* p. 356. See above.

6 2009. *Ocean Friendly Gardens: A How-To Gardening Guide to Help Restore a Healthy Coast and Ocean.* The Surfrider Foundation. See above.

7 1995. *Environmental Hydrology.* See above.

8 2001. Dines, Nicholas and Brown, Kyle. *Landscape Architect's Portable Handbook.* McGraw Hill, 2001, p. 213.

9 *Virginia Erosion and Sediment Control Handbook.* "Minimum Standard 3.10: General Infiltration Practices." Virginia Department of Environmental Quality. https://www.deq.virginia.gov/water/storm-water/stormwater-construction/handbooks Accessed March 18, 2023. Fairly universal, soils with infiltration rates less than .52" per hour or greater than 8.75" must be avoided.

10 1995. *Environmental Hydrology.* p. 356. See above.

11 Lugwid, Art. *Water Storage: Tanks, Cisterns, Aquifers, and Ponds for Domestic Supply, Fire and Emergency Use—Includes How to Make Ferrocement Water Tanks,* 2nd ed. Oasis Design. 2005.
    and
    Banks, Suzy and Heinichen, Richard. *Rainwater Collection for the Mechanically Challenged.* Tank Town Publishing. 2004.

12 Douglas Kent and Mathew West. *Rainwater Harvesting at the Lyle Center for Regenerative Studies, an Examination of the Ecological and Economic Costs and Benefits,* presented to the School of Environmental Design. 2005. We performed an audit for Cal Poly Pomona's Center for Regenerative Studies (CA) and we calculated that the economic and energy costs to capture the potential 70,000 gallons of rainwater falling on the roofs every year. Installing and maintaining the infrastructure needed to capture, transport, screen, store and distribute the rainwater cost 42 BTUs per gallon; the water imported to the center cost 17.3 BTUs per gallon The financial costs were just as divergent; catching and cleaning water onsite cost $1.90 per 100 gallons, whereas the municipal supply cost ¢.25, a 760% difference.

13 2004. *Rainwater Collection for the Mechanically Challenged.* Tank Town Publishing.

14 Ward, Andy D. and Trimble, Stanley W. *Environmental Hydrology,* 2nd ed. CRC Press, 1995, p. 356.

Index of Terms

Index of Common Names for Plants

Index of Botanical Names for Plants